BestMasters

T0238444

Springer awards „BestMasters" to the best master's theses which have been completed at renowned universities in Germany, Austria, and Switzerland.

The studies received highest marks and were recommended for publication by supervisors. They address current issues from various fields of research in natural sciences, psychology, technology, and economics.

The series addresses practitioners as well as scientists and, in particular, offers guidance for early stage researchers.

Paul M. Geffert

Stochastic Non-Excitable Systems with Time Delay

Modulation of Noise Effects by Time-Delayed Feedback

Foreword by Prof. Dr. Eckehard Schöll, PhD

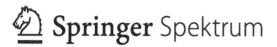 Springer Spektrum

Paul M. Geffert
London, United Kingdom

BestMasters
ISBN 978-3-658-09294-8   ISBN 978-3-658-09295-5 (eBook)
DOI 10.1007/978-3-658-09295-5

Library of Congress Control Number: 2015933640

Springer Spektrum

Springer Spektrum is a brand of Springer Fachmedien Wiesbaden
Springer Fachmedien Wiesbaden is part of Springer Science+Business Media
(www.springer.com)

Foreword

Stochastic effects in nonlinear dynamical systems are an important field of state-of-the-art research. The interplay of noise, nonlinearity, and time-delayed feedback leads to a wealth of novel, unexpected phenomena, such as stochastic bifurcations, coherence resonance, etc. Stochastic bifurcation denotes the transition from a monomodal to a bimodal stationary probability distribution, and coherence resonance is a counterintuitive effect which describes the nonmonotonic dependence of the coherence of noise-induced oscillations upon noise strength, resulting in an optimum coherence at non-zero noise strength.

This Master Thesis focusses on these effects in a simple paradigmatic model, i.e., the Stuart-Landau oscillator. The model variant which is considered in this thesis arises from the generic expansion of an oscillator system near a subcritical Hopf bifurcation in terms of a fifth-order polynomial in the complex variable z. While the third-order, supercritical form of the Stuart-Landau oscillator has been well studied, much less is known about the effect of noise near a subcritical Hopf bifurcation which creates an unstable limit cycle in the deterministic system. This thesis considers the regime of bistability between the trivial zero steady state and a stable limit cycle which arises from the saddle-node bifurcation of the unstable limit cycle. Originally, the term coherence resonance has been restricted to excitable systems, but this thesis extends the notion to non-excitable systems with a subcritical Hopf bifurcation, where coherence resonance is due to a novel mechanism.

The thesis demonstrates that coherence resonance can be modulated by time-delayed feedback or coupling, i.e., enhanced or suppressed depending upon the delay time. Analytical methods are developed and detailed numerical simulations for various ranges of the noise intensity, bifurcation parameter, delay time and feedback strength are presented for a single and for two coupled systems.

This thesis presents novel results on the interplay of noise and delay in nonlinear systems, and in particular offers new insights into the modulation of coherence resonance by time-delayed feedback.

Eckehard Schöll

Profile of the Institute

The group of Prof. Eckehard Schöll at the Institut für Theoretische Physik of the Technische Universität Berlin has long-standing experience in nonlinear dynamics and control. In the center of the activities of the group have been theoretical investigations and computer simulations of nonlinear dynamical systems and complex networks with a particular emphasis on self-organized spatio-temporal pattern formation and its control by time-delayed feedback methods, and stochastic influences and noise. Recent research has focused on the deliberate control and selection of complex, chaotic, or noise-induced space-time patterns as well as synchronization and dynamics of complex delay-coupled networks, and chimera states. As state-of-the-art applications optoelectronic and neural systems are investigated, in particular coupled semiconductor lasers and optical amplifiers, nonlinear dynamics in semiconductor nanostructures, quantum dot lasers with optical feedback and injection, and neuronal network dynamics. The group is active in many national and international collaborations, and in particular is strongly involved in the Collaborative Research Center SFB 910 on Control of Self-Organizing Nonlinear Systems (Coordinator: Eckehard Schöll) and SFB 787 on Semiconductor Nanophotonics of the Deutsche Forschungsgemeinschaft (DFG).

Acknowledgements

First of all, I would like to thank Prof. Eckehard Schöll for suggesting to me this interesting topic and the possibility to write the thesis under his supervision.

I am indebted to Andrea Vüllings for her support during the whole process of writing this thesis and for sharing an office. She pointed out any ambiguities and gave new input and ideas, when I felt lost or the progress was stuck.

Further thanks go to Anna Zakharova for giving useful hints and comments concerning stochastic bifurcations. I thank the whole Schöll group for a friendly and comfortable atmosphere, also besides the work.

Furthermore, I appreciate the discussions and the fruitful collaboration with Wolfram Just.

Last but, not least, I want to thank my family, especially my parents, for their never-ending support from every point of view.

This work was written within the framework of SFB 910.

Part of this work was published as [63] and is reused with kind permission of The European Physics Journal (EPJ) and Springer Science+Business Media.

Paul M. Geffert

Contents

List of Figures

Chapter 1

Introduction

Random perturbations in dynamical systems are an ubiquitous phenomenon in many fields of science. These perturbations, e.g. represented by noise, lead to an unpredictable movement of the trajectories of the system. It is of great interest to control such a random motion or other noise-induced effects.

Adding a random force term to the deterministic equations of a dynamical system opens a new field of research with many new phenomena [1, 2]. New concepts for the investigations have been developed in order to study stochastic systems, e.g. a rigorous mathematical theory of stochastic bifurcations [3]. Noise is usually unwanted because of the loss of predictability for the trajectory of a system. However, contrary to what one would expect, it turns out that noise can also play a constructive role in nonlinear systems. Some famous and well studied examples are the stochastic resonance [4] and coherence resonance [5–7]. Stochastic resonance describes the effect that a weak periodic force, which drives a stochastic system, is enhanced at an intermediate noise strength. Coherence resonance denotes the maximum regularity of noise-induced oscillations without an external driving force at a non-zero noise strength.

The investigation of delay differential equations has become a central topic in current research. Time delay is a proper description for many processes in science. Some examples from very different fields are presented in [8], e.g. population dynamics, chemical reactions, or lasers. A delay differential equation increases the dimension of the phase space to infinity [9], in contrast to an ordinary differential equation, so the initial condition has to be specified in a time interval ("history function") rather than at a single time (single initial value). Analytical studies are hampered as a result.

Time-delayed feedback control [10] was suggested as an improved concept for chaos control [11, 12]. This method (also known as Pyragas control) has become very popular and has been applied in many different areas of research: stabilisation of unstable fixed points and periodic orbits [13–16], various applications to experimental setups [17–19], mathematical theorems [20–22], even in fluid dynamics [23] and quantum systems [24]. Also in the context of controlling nonlinear stochastic systems time-delayed feedback control has been applied successfully [25, 26].

The studies of networks has attracted large attention because of the potential application in many fields of research [27, 28]. Coupling terms with time delay are a convenient way to describe for example the finite signal transmission between nodes of the network. The type of coupling or the suitable choice of the coupling parameters plays an important role in studies on synchronisation [29–34], in connection with chaos [35, 36] or with the aim to stabilize periodic orbits [37, 38]. The influence of delayed-coupling in systems with noise was investigated for neuronal systems [39–42] and coupled lasers [43–45].

In this thesis, we want to investigate the interplay between noise and time-delay in Hopf normal forms (Stuart-Landau oscillators). More precisely, we are interested in the interplay or the possible modulation of the dynamical system by time-delayed coupling and noise.
In chapter 2, a single Hopf normal form with a random force term is investigated. We start with the bifurcation and linear stability analysis of the deterministic model. Then we add noise to the system and apply statistical linearisation techniques to obtain analytical expressions for the study of the noise effects. To conclude this chapter, we suggest a new quantity to uncover the mechanism of coherence resonance in non-excitable systems.
Chapter 3 contains the investigations on the stochastic Hopf normal form with time-delayed feedback. As in chapter 2, we start with a deterministic bifurcation and stability analysis. Then we consider the stochastic time-delayed system. A multiple scale perturbation expansion is developed to obtain an analytical expression for the stationary amplitude probability distribution. The modulation of coherence resonance and its mechanism in the presence of time delay is discussed.
In chapter 4 we will try to extend the methods, developed in the chapters 2 and 3, to coupled Hopf normal forms for the investigations of noise effects.
Chapter 5 gives a summary of this thesis and chapter 6 provides an outlook, where ideas and literature is suggested for the investigation of the noise effects in network motifs or even large networks of coupled Hopf normal forms, which is beyond the scope of this thesis.

Chapter 2

Stochastic effects in nonlinear systems

Oscillations are a frequently studied phenomenon in science. Examples of this widespread behaviour range from economy to natural sciences. Neurons also exhibit oscillatory properties. The Hodgkin-Huxley model [46] and the FitzHugh-Nagumo model [47, 48] belong to the most famous models to describe the dynamics of neurons. The FitzHugh-Nagumo model is used more widely, because the calculations for large networks of neurons are easier to handle with this system. It is also the prototype of an excitable system [49].

Excitable systems can be described by three different states: a rest state, an excited state and a refractory phase. If the system is perturbed sufficiently above a certain threshold, the trajectory is kicked out of a locally stable fixed point (rest state) and makes a long excursion through the phase space emitting a spike (excited state). After the excursion the trajectory moves back to the stable fixed point (refractory phase). Such a perturbation can be realised by noise. Noise-induced oscillations can be found in excitable system (in the excitable regime), as already mentioned, but also in system close to bifurcations (non-excitable systems). The latter will be considered in this work.

A famous example for a non-excitable system is the Van der Pol model [50], which is also the prototype a nonlinear oscillator. The behaviour of this system can be divided into two parameter regions: in one parameter region the oscillator shows damped oscillations, whereas the other regime is characterised by self-sustained oscillations. To change from one regime to the other, the system undergoes a Hopf bifurcation. A generic model to describe the behaviour of non-excitable systems close to a Hopf bifurcation is the Hopf normal form, which is also known under the name Stuart-Landau oscillator.

In this thesis, a Hopf normal form with a random force and time-delayed coupling term is studied. Before we investigate the full stochastic and time-delayed dynamics, we start with a single oscillator system with noise in this chapter. First, we will understand the deterministic behaviour of our system.

2.1   Deterministic dynamics

The deterministic equation of motion reads

$$\dot{z}(t) = (\lambda + i\omega_0 - a|z(t)|^2 - b|z(t)|^4)z(t). \tag{2.1}$$

$z(t)$ is a complex variable, λ denotes the bifurcation parameter, and ω_0 is the intrinsic frequency of the system. The real parameters a and b are used to distinguish between the supercritical ($a = 1$, $b = 0$) and the subcritical ($a = -1$, $b = 1$) Hopf normal form.
The complex variable z can be expressed in polar coordinates, $z = re^{i\phi}$ ($r \geq 0$), and decomposed into its real and imaginary part. We obtain the following equations.

$$\dot{r} = \lambda r - ar^3 - br^5, \tag{2.2}$$
$$\dot{\phi} = \omega_0. \tag{2.3}$$

For the bifurcation diagram of the supercritical Hopf normal form, we set $a = 1$ and $b = 0$. The stationary solution or fixed point ($\dot{r} = 0$) of Eq. (2.2) is given by

$$r_1^* = 0. \tag{2.4}$$

For $\lambda > 0$ there exists a periodic solution (limit cycle)

$$z_2 = r_2^* e^{i\omega t} \text{ with } r_2^* = \sqrt{\lambda}. \tag{2.5}$$

For the subcritical Hopf normal form, we set $a = -1$ and $b = 1$. Then the fixed point is

$$r_1^* = 0, \tag{2.6}$$

and the periodic solutions are

$$r_2^* = \sqrt{\frac{1 + \sqrt{1 + 4\lambda}}{2}}, \ r_3^* = \sqrt{\frac{1 - \sqrt{1 + 4\lambda}}{2}}. \tag{2.7}$$

Now we have to investigate the stability of the fixed point r^* and the limit cycles and, therefore, we make a linear stability analysis [51].
We consider a dynamical system described by the differential equation $\dot{x} = f(x)$, where x represents the dynamical variable and $f(x)$ is some nonlinear function of this variable. This system will be linearised in the vicinity of its fixed point x^*

$$\delta x = x - x^*, \tag{2.8}$$

where δx denotes a small deviation. We are interested in the time evolution of these deviations, so we derive a differential equation

$$\delta \dot{x} = \dot{x} - \dot{x}^* = f(x) - f(x^*) = f(x), \tag{2.9}$$

where we used $\dot{x}^* = f(x^*) = f(x)|_{x=x^*} = 0$. We can rewrite $f(x) = f(x^* + \delta x)$ and perform a Taylor series expansion for small deviations δx around the fixed point x^* and we find

$$\delta \dot{x} = f(x^*) + \frac{d}{dx}f(x)\bigg|_{x=x^*} \delta x + \mathcal{O}(\delta x^2)$$

$$\approx \frac{d}{dx}f(x)\bigg|_{x=x^*} \delta x. \tag{2.10}$$

Equation (2.10) is a differential equation for the behaviour of a system in the vicinity of a fixed point. This problem can be solved using an exponential ansatz for the deviations $\delta x \propto e^{\Lambda t}$. More generally, in n dimensions, $\frac{d}{dx}f(x)\big|_{x=x^*}$ is replaced by the Jacobian matrix Df, so we have to deal with an eigenvalue problem of the Jacobian.

For the fixed point r_1^* (Eqs. (2.4, 2.6)) we find that it is stable for $\lambda < 0$ and becomes unstable for $\lambda > 0$.

For periodic orbits, such as limit cycles, Floquet theory answers the question as to whether a periodic state is stable or unstable, because in this case the Jacobian matrix is in general time-dependent. For the Stuart-Landau oscillator this time dependence of the Jacobian matrix vanishes, so it is possible to perform the linear stability analysis analytically.

We obtain for the supercritical case

$$\begin{pmatrix} \delta \dot{r} \\ \delta \dot{\phi} \end{pmatrix} = \begin{pmatrix} \lambda - 3r^2 & 0 \\ 0 & 0 \end{pmatrix}\bigg|_{r^*} \begin{pmatrix} \delta r \\ \delta \phi \end{pmatrix} = \underbrace{\begin{pmatrix} -2\lambda & 0 \\ 0 & 0 \end{pmatrix}}_{M} \begin{pmatrix} \delta r \\ \delta \phi \end{pmatrix}. \tag{2.11}$$

The eigenvalues of the Matrix M are the Floquet exponents Λ:

$$\Rightarrow \Lambda = \begin{cases} 0 & \text{Goldstone mode (invariance of translation along the limit cycle)} \\ -2\lambda \end{cases}$$

$$\tag{2.12}$$

Hence the limit cycle, which exists for $\lambda > 0$ (see Eq. (2.5)), is stable.

For the subcritical Hopf normal form we have

$$\begin{pmatrix} \delta \dot{r} \\ \delta \dot{\phi} \end{pmatrix} = \begin{pmatrix} \lambda + 3r^2 - 5r^4 & 0 \\ 0 & 0 \end{pmatrix}\bigg|_{r^*} \begin{pmatrix} \delta r \\ \delta \phi \end{pmatrix}$$

$$= \underbrace{\begin{pmatrix} \lambda + 3\left(\frac{1\pm\sqrt{1+4\lambda}}{2}\right) - 5\left(\frac{1\pm\sqrt{1+4\lambda}}{2}\right)^2 & 0 \\ 0 & 0 \end{pmatrix}}_{M} \begin{pmatrix} \delta r \\ \delta \phi \end{pmatrix}, \tag{2.13}$$

$$\Rightarrow \Lambda = \begin{cases} 0 & \text{Goldstone mode} \\ \lambda + 3\left(\frac{1\pm\sqrt{1+4\lambda}}{2}\right) - 5\left(\frac{1\pm\sqrt{1+4\lambda}}{2}\right)^2 \end{cases} \tag{2.14}$$

The limit cycle with radius r_3^* (Eq. (2.7)) exists for $\lambda \in [-\frac{1}{4}, 0]$, and is unstable because the Floquet exponent

$$\Lambda = \lambda + 3 \left(\frac{1 - \sqrt{1 + 4\lambda}}{2} \right) - 5 \left(\frac{1 - \sqrt{1 + 4\lambda}}{2} \right)^2 = -1 - 4\lambda + \sqrt{1 + 4\lambda} \quad (2.15)$$

is always positive for $-\frac{1}{4} < \lambda < 0$ since

$$\begin{aligned} & 1 + 4\lambda > (1 + 4\lambda)^2 \\ \Leftrightarrow \quad & 4 < 8 + 16\lambda \\ \Leftrightarrow \quad & 0 < 1 + 4\lambda \quad (\lambda < 0). \end{aligned} \quad (2.16)$$

However, the limit cycle with radius r_2^* (Eq. (2.7)), which exists for $\lambda \in [-\frac{1}{4}, \infty]$, is stable because the corresponding Floquet exponent

$$\Lambda = \lambda + 3 \left(\frac{1 + \sqrt{1 + 4\lambda}}{2} \right) - 5 \left(\frac{1 + \sqrt{1 + 4\lambda}}{2} \right)^2 = -1 - 4\lambda - \sqrt{1 + 4\lambda} \quad (2.17)$$

is negative for $\lambda > -\frac{1}{4}$.

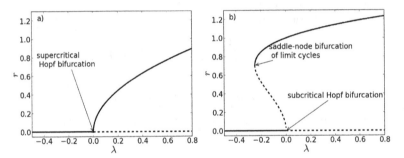

Figure 2.1: Deterministic bifurcation diagram; solid lines correspond to stable focus/limit cycle, dashed lines to unstable ones.
(a) supercritical case ($a = 1, b = 0$), see Eqs. (2.4, 2.5, 2.12),
(b) subcritical case ($a = -1, b = 1$), see Eqs. (2.6, 2.7, 2.15, 2.17).

In Fig. 2.1a) for $\lambda < 0$ we have a stable focus, which loses its stability via a supercritical Hopf bifurcation at $\lambda = 0$. For $\lambda > 0$ a stable limit cycle and an unstable focus coexist. In Fig. 2.1b) we also have a stable focus for $\lambda < 0$, which undergoes a subcritical Hopf bifurcation at $\lambda = 0$. This Hopf bifurcation gives rise to an unstable limit cycle. At $\lambda = -0.25$ a saddle-node bifurcation of limit cycles takes place and a stable limit cycle is born. The system is bistable for values of $\lambda \in [-0.25, 0]$.

2.2   Stochastic dynamics

An additional noise term in the dynamical equations leads to many new dynamical features [1, 49], e.g. noise-induced oscillations. For example, a system in the regime of a deterministically stable focus is influenced by noise in such a way that the trajectory is kicked out of the focus and shows random motion around it. Adding a random force term to Eq. (2.1), we obtain

$$\dot{z}(t) = (\lambda + i\omega_0 - a|z(t)|^2 - b|z(t)|^4)z(t) + \sqrt{2D}\xi(t). \tag{2.18}$$

$D \geq 0$ describes the strength of the fluctuations (noise strength), whereas $\xi(t) \in \mathbb{C}$ denotes the random variable. Here Gaussian white noise is used with the following properties

$$\langle \xi(t) \rangle = 0, \quad \langle \xi(t)\xi^*(t') \rangle = 2\delta(t - t'). \tag{2.19}$$

The white noise is usually used for describing an unknown perturbation, where all frequencies contribute equally. This can be shown via Fourier transform (Wiener-Khinchin theorem), where the power spectral density is a constant function [53]. For the investigations of stochastic systems we have to adapt our methods, because we are dealing with random variables. Their properties can be described by probability distributions. Therefore, the deterministic concepts like bifurcations or attractors cannot be applied, or have to be modified (see, e.g. stochastic bifurcations).

To calculate the probability distribution for our dynamical system (Eq. (2.18)), we derive the corresponding Fokker-Planck equation. This computation can be found in many textbooks, e.g. [53–55]. The Fokker-Planck equation for the amplitude r and phase ϕ (z is decomposed in polar coordinates) corresponding to Eq. (2.18) is:

$$\partial_t P = \partial_r \left(\left(\left(-\lambda r + ar^3 + br^5 - \frac{D}{r} \right) P + D\partial_r P \right) + \partial_\phi \left(-\omega_0 P + \frac{D}{r^2}\partial_\phi P \right) \right. \tag{2.20}$$

where $P(r, \phi)$ is the probability density. A detailed derivation of Eq. (2.20) is given in the Appendix A.

By using the spherical symmetry of the deterministic system, we neglect the derivatives with respect to the phase variable ($\partial_\phi P = 0$). Furthermore, we are only interested in the stationary behaviour of our system, therefore, we have $\partial_t P = 0$. We obtain the stationary amplitude probability distribution

$$P(r) = Nr \exp \left(\frac{r^2}{D} \left(\frac{\lambda}{2} - \frac{ar^2}{4} - \frac{br^4}{6} \right) \right), \tag{2.21}$$

where N is the normalisation constant and given by

$$N = \left(\int_0^\infty r \exp \left(\frac{r^2}{D} \left(\frac{\lambda}{2} - \frac{ar^2}{4} - \frac{br^4}{6} \right) \right) dr \right)^{-1}. \tag{2.22}$$

2.3 Stochastic bifurcations

In deterministic systems, a bifurcation is a sudden change of the behaviour of the system after varying a certain system parameter, which is the so-called bifurcation parameter. An example was shown in the beginning of this chapter, where λ is the bifurcation parameter and the system undergoes a Hopf bifurcation, which means a change from damped oscillations to self-sustained oscillations (Eq. (2.1) and Fig. 2.1).

In a system with noise (stochastic system), the deterministic concept of bifurcation theory cannot be directly applied but must be adapted. Here the noise intensity D works as the control- or bifurcation parameter. Following [3] there are two types of stochastic bifurcations: the phenomenological bifurcation (P-bifurcation) denotes a change in the shape of the probability distribution, e.g. from unimodal to a bimodal shape. The other type is the dynamical bifurcation (D-bifurcation) which occurs, when the largest Lyapunov exponent becomes positive as a function of the noise intensity. The latter will not be considered in this work.

There are many applications and investigations on stochastic bifurcations and noise-induced transitions: studies on generic models [56, 57], biological and chemical systems [58, 59] and experimental and theoretical investigations in lasers [60]. Many further examples can be found in [2]. For a generalised Van der Pol model, the stochastic P-bifurcation was investigated in [61, 62].

First, we will investigate the supercritical case ($a = 1$, $b = 0$) for our system: according to Eq. (2.21) the probability distribution is given by

$$P(r) = N r \exp\left(-\frac{W(r)}{D}\right) \tag{2.23}$$

where $W(r)$ is the potential

$$W(r) = \frac{r^4}{4} - \frac{\lambda r^2}{2}. \tag{2.24}$$

Here we follow [61, 62] for further steps of investigation. We are looking for the extrema of the distribution

$$\frac{d}{dr}P(r) = N \exp\left(-\frac{W(r)}{D}\right)\left(1 - \frac{r}{D}\frac{d}{dr}W(r)\right) = 0. \tag{2.25}$$

A stochastic P-bifurcation takes place if the number of extrema changes. The relevant part from Eq. (2.25) reads

$$1 - \frac{r}{D}\frac{d}{dr}W(r) = 1 - \frac{r^4}{D} + \frac{\lambda r^2}{D}. \tag{2.26}$$

The extrema of the distribution are the real-valued, positive roots of Eq. (2.26). It is

$$D - r^4 + \lambda r^2 = 0 \quad \rightarrow \quad f(x) = x^2 - \lambda x - D = 0 \tag{2.27}$$

with the substitution $x = r^2$. The number of extrema changes if $f(x_c) = f'(x_c) = 0$, where x_c denotes the corresponding extremum. So $f'(x)$ reads

$$2x - \lambda = 0, \quad \to \quad x_c = \frac{\lambda}{2} \tag{2.28}$$

Inserting this result into Eq. (2.27), we find that

$$D = -\frac{\lambda^2}{4}. \tag{2.29}$$

Equation (2.29) shows that a transition will only take place for negative noise intensities, but D has to be positive. That implies that a stochastic P-bifurcation of the amplitude probability distribution cannot occur in the supercritical case. This is shown in Fig. 2.2, where the number of the maxima does not change by increasing the noise intensity.

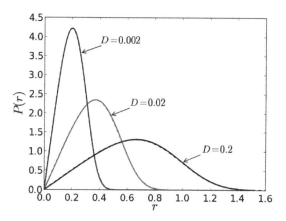

Figure 2.2: Stationary amplitude probability distribution for the super-critical case, analytically (solid, according to Eqs. (2.23, 2.24)) and numerically (dashed) calculated. Parameters: $\omega_0 = 2\pi$, $\lambda = -0.01$.

For the subcritical normal form ($a = -1$, $b = 1$) we follow [63] for another way of the calculation. The probability distribution reads

$$P(r) = Nr \exp\left(\frac{r^2}{D}\left(\frac{\lambda}{2} + \frac{r^2}{4} - \frac{r^4}{6}\right)\right), \tag{2.30}$$

where N represents the corresponding normalisation constant (see Eq. (2.22)). We present the detailed calculation for the bifurcation lines in the Appendix B:

a condition for the variable r^2 is found

$$r^2 = -\frac{9D + \lambda}{6\lambda + 2} > 0 \text{ and } r \in \mathbb{R}. \tag{2.31}$$

Equation (2.31) shows that only values of $\lambda < 0$ satisfy the condition. The bifurcation lines are represented by

$$D_{1,2} = \frac{1}{27}\left(-9\lambda - 2\left(1 \pm \sqrt{(1 + 3\lambda)^3}\right)\right). \tag{2.32}$$

This result can be seen in the following stochastic bifurcation diagram.

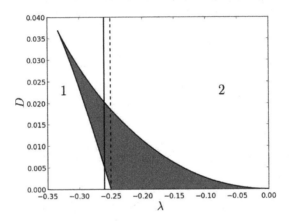

Figure 2.3: Stochastic bifurcation diagram following from Eq.(2.32) for the subcritical case ($a = -1, b = 1$). The dashed line corresponds to the boundary between deterministic monostability (region 1) and bistability (region 2). The solid line denotes the parameter line for the fixed deterministic bifurcation parameter λ. The shaded region denotes the parameter regime where the distribution has a bimodal shape. Outside of this region we have a unimodal distribution.

Figure (2.3) shows the stochastic bifurcation diagram for our noisy system: the shaded region denotes the parameter values, where the amplitude probability distribution has a bimodal shape. The dashed vertical line describes the border between the two different deterministic regimes corresponding to λ: in the left part (region 1) there is only a stable focus, whereas in the right part (region 2) the deterministic system is bistable, because additionally a stable limit cycle coexists with the stable focus. The vertical solid line denotes the λ value of interest; for the further investigations, $\lambda = -0.26$ is fixed, and we vary the noise intensity D. By crossing the solid lines to enter or to leave the shaded region (parameter regime of

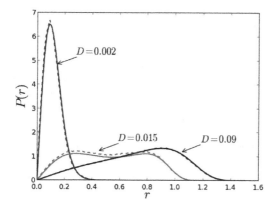

Figure 2.4: Amplitude probability distribution (Eq. (2.30)) for the subcritical case for different noise strengths, analytically (solid)
and numerically (dashed) calculated.
Parameters: $\lambda = -0.26$, $\omega_0 = 2\pi$.

bimodality), we observe a stochastic P-bifurcation.
Figure 2.4 shows the probability distribution for values of the noise intensity D
in and outside from the region of bimodality. Furthermore, the analytical result
and the numerics for the amplitude probability distribution are shown: they are
in excellent agreement.
The discussion above was carried out for additive white noise sources. An investigation for coloured noise, additive and multiplicative, can be found in [64], where
a Duffing-van der Pol model is investigated.

2.4 Coherence resonance

The presence of noise is usually unwanted because of its destructive character for
the deterministic dynamic. Nevertheless, noise can have a constructive effect in the
interplay with the nonlinearities of a dynamical system. A very well studied effect
is the stochastic resonance [4]: a weak periodic signal, which drives the dynamical
system, is enhanced by the noise. The counterintuitive aspect here is that the best
enhancement takes place at an intermediate non-zero noise strength.
Another phenomenon that is also related with the constructive role of the noise is
coherence resonance. In contrast to stochastic resonance, there is no external periodic force present. This effect occurs from the intrinsic dynamic of the system; the
noise-induced oscillations become most regular at a finite non-zero noise strength.
Coherence resonance was investigated first by Gang et al. [5], who studied this phenomenon in a generic model for a SNIPER (saddle-node infinite period) bifurcation

(type I excitability). But their work was published with the title "Stochastic resonance without external periodic force". The catchy name "coherence resonance" was invented by Pikovsky and Kurths, who investigated the FitzHugh-Nagumo model (type II excitability), [6].

Coherence resonance was also investigated in a non-excitable system in [7, 61, 62]. [7] was a combined work with a laser experiment and theoretical modelling. In the experiment, the authors investigated resonance effects caused by noise, setting the laser close to a Hopf bifurcation. The theory was based on the estimate of a Lorentzian spectrum. This suggestion was provided by the experiment and numerical simulations of a noisy Hopf normal form. With the help of the Wiener-Khinchin theorem, the authors developed an amplitude approach to calculate measures of coherence (see section 2.4.1). They found that coherence resonance occurs only in the subcritical case. Therefore, we will not consider the supercritical case in the following investigations.

Coherence resonance can be observed also below other types of bifurcations, where characteristic signatures of noisy precursors occur [65, 66], in lasers [7, 45, 67, 68], and neural systems [6, 39, 49, 69], just to mention a few examples. We will now investigate the same theoretical model as used in [7], but our starting point for the calculations is based on the generic Eq. (2.18). Using statistical linearisation techniques, we are able to derive analytical expressions for the measures of coherence. This measures will be introduced in the following section.

2.4.1 Measures of coherence

Usually, the following three measures are used to quantify the degree of regularity of the oscillations:

- The **signal-to-noise-ratio** (SNR)

$$\beta = \frac{H}{\Delta\omega/\omega_p}. \tag{2.33}$$

  β is calculated from the power spectral density (see Fig. 2.5) and is usually used in laser physics. It was introduced by Haken, see [5].

- The normalised fluctuations of the **interspike interval** (ISI)

$$R = \frac{\sqrt{\langle t_p^2 \rangle - \langle t_p \rangle^2}}{\langle t_p \rangle}. \tag{2.34}$$

  In excitable systems, it is useful to look at the time series and to determine the statistics of spikes [6]. For a non-excitable system, this measure is not used, because there are no spikes in the sense of an excitation.

- The **correlation time**

$$t_{cor} = \frac{1}{\Psi(0)} \int_0^\infty |\Psi(s)| ds, \tag{2.35}$$

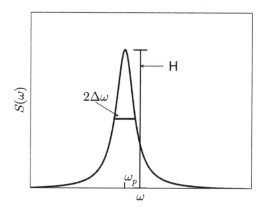

Figure 2.5: Example of a power spectral density.

which is calculated from the autocorrelation function

$$\Psi(s) = \langle [x(t+s) - \langle x \rangle][x(t) - \langle x \rangle] \rangle. \tag{2.36}$$

$\Psi(0) = \sigma^2$ is the variance and $\langle \ldots \rangle$ denotes the ensemble average. This measure was introduced by Stratonovich [70] and shows, how fast the correlation decays in a stochastic process. For a linear stochastic process the autocorrelation function is of the form $\Psi(s) \approx \exp(-\frac{2}{\pi}\frac{s}{t_{cor}})\cos(\omega s)$, see Fig. 2.6 [52, 77].

These measures are usually plotted as a function of the noise strength. If these curves have an extremum, the system will show (in nearly all cases, see [7] for an exception) coherence resonance. Now we want to give analytic expressions of these measures for our system (Eq. (2.18)). To reach this goal we have to simplify our nonlinear Eq. (2.18) by using suitable methods.

We will use the concept from [71] which is known as statistical linearisation. The nonlinear terms from Eq. (2.18) (subcritical case $a = -1$, $b = 1$) $|z(t)|^2 z(t) - |z(t)|^4 z(t)$ are replaced by an effective linear term $\alpha z(t)$ [63], where α is a constant. This effective term is estimated under the condition that the deviations due to the nonlinearities are minimized

$$\left\langle \left| \left(|z(t)|^2 z(t) - |z(t)|^4 z(t) - \alpha z(t) \right) \right|^2 \right\rangle \to \min. \tag{2.37}$$

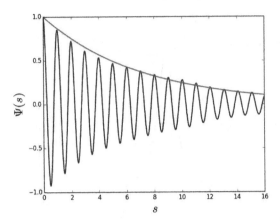

Figure 2.6: Autocorrelation function and the exponential decaying envelope, numerically calculated (Eq. (2.18)).
Parameters: $a = -1, b = 1, \omega_0 = 2\pi, D = 0.05, \lambda = -0.26$.

where $\langle \ldots \rangle = \int_0^\infty \ldots P(r)dr$ denotes the expectation value due to the stationary amplitude probability distribution (Eq. (2.30)). Variation with respect to α gives

$$
\begin{aligned}
0 =& \frac{d}{d\alpha} \left\langle \left| \left(|z(t)|^2 z(t) - |z(t)|^4 z(t) - \alpha z(t) \right) \right|^2 \right\rangle \\
=& \left\langle -z^*(t) \left(|z(t)|^2 z(t) - |z(t)|^4 z(t) - \alpha z(t) \right) - cc. \right\rangle \\
=& \left\langle -2(|z(t)|^4 - |z(t)|^6 - \alpha |z(t)|^2) \right\rangle .
\end{aligned}
\tag{2.38}
$$

Thus, we obtain for the effective coefficient

$$
\alpha = \frac{\langle |z(t)|^4 \rangle - \langle |z(t)|^6 \rangle}{\langle |z(t)|^2 \rangle}.
\tag{2.39}
$$

Other approximations are used in this way to replace the nonlinearities by their mean value, assuming a Gaussian distribution [52, 62, 72]. In the context of such a self-consistent mean-field approximation, it is possible to determine the mean value self-consistently and also analytically, whereas the coefficient α (Eq. 2.39) has to be determined numerically. The self-consistent mean-field calculations captures the nonlinear effects of the underlying system quite well [52], but are also limited to certain parameter values and only show qualitative agreement with the numerics when nonlinear terms of high order contribute to the dynamics, because the underlying distribution is no longer a Gaussian distribution [73].

After the statistical linearisation of Eq. (2.18), we obtain a linear stochastic process (Ornstein-Uhlenbeck process). Now we can compute the power spectral density and the measures of coherence analytically [53]:

$$\dot{z}(t) = (\tilde{\lambda} + i\omega_0)z(t) + \sqrt{2D}\xi(t), \tag{2.40}$$

where $\tilde{\lambda} = \lambda + \alpha$ denotes the effective system parameter. We decompose Eq. (2.40) into real and imaginary parts ($z = x + iy$) and obtain a two-dimensional system:

$$\begin{pmatrix} dx \\ dy \end{pmatrix} = -\underbrace{\begin{pmatrix} -\tilde{\lambda} & \omega_0 \\ -\omega_0 & -\tilde{\lambda} \end{pmatrix}}_{=:\,\mathbf{A}} \begin{pmatrix} x \\ y \end{pmatrix} dt + \underbrace{\begin{pmatrix} \sqrt{2D} & 0 \\ 0 & \sqrt{2D} \end{pmatrix}}_{=:\,\mathbf{B}} \begin{pmatrix} dW_1 \\ dW_2 \end{pmatrix}. \tag{2.41}$$

We compute the power spectral density

$$\mathbf{S}(\omega) = (\mathbf{A} + i\omega)^{-1}\mathbf{B}\mathbf{B}^T(\mathbf{A}^T - i\omega)^{-1}. \tag{2.42}$$

For the x-variable, we find

$$S_{xx}(\omega) = D\left(\frac{1}{(\omega - \omega_0)^2 + \tilde{\lambda}^2} + \frac{1}{(\omega + \omega_0)^2 + \tilde{\lambda}^2}\right). \tag{2.43}$$

This gives us expressions for the peak height H, the full width at half maximum $2\Delta\omega$ and, therefore, also for the signal-to-noise ratio β, which are:

$$H = \frac{D}{\tilde{\lambda}^2}, \quad 2\Delta\omega = -2\tilde{\lambda}, \quad \tilde{\lambda} < 0, \tag{2.44}$$

$$\rightarrow \quad \beta = \frac{H}{\Delta\omega} = -\frac{D}{\tilde{\lambda}^3}. \tag{2.45}$$

The autocorrelation function can be calculated via the Wiener-Khinchin theorem from the power spectral density (Eq. (2.43)). Here it is calculated from the solution of Eq. (2.41) [53]:

$$\mathbf{X}(t) = \exp(-\mathbf{A}t)\mathbf{X}(0) + \int_0^t \exp[-\mathbf{A}(t - t')]\mathbf{B}\,\xi(t')dt'. \tag{2.46}$$

Using the stationary solution ($t \to \infty$)

$$\mathbf{X}_s(t) = \int_{-\infty}^t \exp[-\mathbf{A}(t - t')]\mathbf{B}\,\xi(t')dt', \tag{2.47}$$

we can calculate the autocorrelation function in the following way

$$\begin{aligned} \boldsymbol{\Psi}(s) &= \left\langle [\mathbf{X}_s(t - s) - \langle\mathbf{X}_s(t - s)\rangle][\mathbf{X}_s(t) - \langle\mathbf{X}_s(t)\rangle]^T\right\rangle \\ &= \langle\mathbf{X}_s(t - s)\mathbf{X}_s(t)^T\rangle, \end{aligned} \tag{2.48}$$

where $\mathbf{X}_s = \begin{pmatrix} x \\ y \end{pmatrix}$ and the zero mean of the stationary solution $\langle\mathbf{X}_s(\tau)\rangle = 0$ (τ represents an arbitrary time variable here).

Therefore, we obtain

$$
\begin{aligned}
\mathbf{\Psi}(\mathbf{s}) &= \left\langle \int_{-\infty}^{t-s} \exp\left[-\mathbf{A}(t-s-t_1)\right]\mathbf{B}\, \xi(t_1)dt_1 \int_{-\infty}^{t} \xi^T(t_2)\, \mathbf{B}^T \exp\left[-\mathbf{A}^T(t-t_2)\right]dt_2 \right\rangle \\
&= \int_{-\infty}^{t-s} \int_{-\infty}^{t} \exp\left[-\mathbf{A}(t-s-t_1)\right]\mathbf{B}\, \underbrace{\left\langle \xi(t_1)\xi^T(t_2) \right\rangle}_{\delta(t_1-t_2)}\, \mathbf{B}^T \exp\left[-\mathbf{A}^T(t-t_2)\right]\, dt_2\, dt_1 \\
&= \int_{-\infty}^{t-s} \exp\left[-\mathbf{A}(t-s-t_1)\right]\mathbf{B}\mathbf{B}^T \exp\left[-\mathbf{A}^T(t-t_1)\right]\, dt_1.
\end{aligned}
\tag{2.49}
$$

For the x component, we thus find

$$
\Psi_{xx}(s) = \frac{-D}{\tilde{\lambda}}e^{\tilde{\lambda}s}\cos(\omega_0 s).
\tag{2.50}
$$

So we can calculate the correlation time

$$
\begin{aligned}
t_{cor} &= \frac{1}{\Psi_{xx}(0)}\int_0^\infty |\Psi_{xx}(s)|ds \\
&= \int_0^\infty \left| e^{\tilde{\lambda}s}\cos(\omega_0 s)\right| ds \\
&= \int_0^\infty e^{\tilde{\lambda}s}\left|\cos(\omega_0 s)\right| ds \\
&\approx \frac{2}{\pi}\int_0^\infty e^{\tilde{\lambda}s}ds \\
&= -\frac{2}{\pi\tilde{\lambda}},
\end{aligned}
\tag{2.51}
$$

where we used the filling factor $\frac{1}{\pi}\int_{-\frac{\pi}{2}}^{\frac{\pi}{2}}\cos(\phi)d\phi = \frac{2}{\pi}$ [52, 77].

The results for the four measures (peak height, width, signal-to-noise ratio, and correlation time) as a function of the noise intensity are shown in Fig. 2.7:
the peak height, the signal to noise ratio and the correlation time exhibit non-monotonic behaviour and a maximum, whereas the width shows a minimum corresponding to the optimal noise intensity at which coherence resonance can be observed. All four measures show excellent agreement with the numerics; the deviations were caused by the statistical linearisation, which minimizes the error between the non-linear and linear equation, but obviously there is still a small difference.

So far, we used the linearisation based on [71]. In [7, 62] a similar approach (amplitude approach) was used, also with a good agreement with the numerics. To calculate the power spectral density and the signal to noise ratio, other methods can also be used. In [74] the calculation is based on a linear-response theory. The analytical results of this powerful technique also show excellent agreement with the numerics.

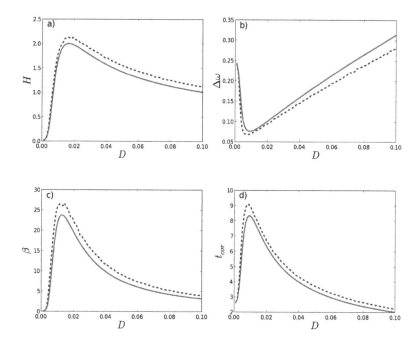

Figure 2.7: Measures of coherence as functions of the noise strength:
a) Peak height H (Eq. (2.44)), b) width at half maximum
$\Delta\omega$ (Eq. (2.44)), c) Signal-to-noise ratio β (Eq. (2.45)), d)
correlation time t_{cor} (Eq. (2.51)). The solid lines correspond
to the analytical calculation and the dashed lines to numerical simulations. Parameters: $\lambda = -0.26$, $\omega_0 = 2\pi$.

2.4.2 Mechanism of coherence resonance

In excitable systems, the mechanism of coherence resonance can be explained by
two characteristic time scales and their different behaviour by increasing the noise.
Here we follow the explanation from [6]: the duration of a spike $t_p = t_a + t_e$ con-
sists of the activation time t_a, which denotes the time to excite the system out
of the fixed point, and the excursion time t_e, which describes the time passing by
during the phase space excursion back to the rest state. For small noise intensities,
the activation time dominates the process, because excitation from the rest state
above the threshold does not occur that often and fluctuates strongly. So we have
$t_a \gg t_e$ and, therefore, $t_p \approx t_a$ with strong variance. For high noise strengths, the
activation time is negligible and we have $t_p \approx t_e$. With increasing noise intensity,
the activation time decreases and the excitation time increases. At an intermediate noise strength the activation time is small and the excursion time is still quite

regular. The system can perform a regular motion, resulting in a minimum of the fluctuations of the pulse duration (see Eq. (2.34)).

For non-excitable systems, we suggest another mechanism, which connects the concept of stochastic bifurcations with coherence resonance [62]. In Fig. 2.8 it can be

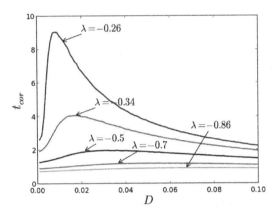

Figure 2.8: Correlation time for different values of λ, numerically calculated (Eq. (2.18)). Parameters: $a = -1, b = 1, \omega_0 = 2\pi$.

observed that the effect of coherence resonance is most pronounced in the regime of a bimodal amplitude probability distribution. The reason is the following: for low noise intensities we only find a distribution with a high peak in the vicinity of the origin: the focus is merely shifted a bit. High noise intensities correspond to a very broadened distribution, the dynamic is smeared out in the phase space. Thus, in both cases one cannot observe regular motion of the trajectory. For intermediate noise intensities, the distribution becomes bimodal. Now there are two preferred regions for the trajectory, so it can perform a regular motion between this two regions. The left peak correspond again to the shifted focus, the right one visualizes the ghost of the limit cycle and as a result the motion becomes most coherent. This leads to coherence resonance.

Numerical simulations and the analytical calculations show that coherence resonance also occurs in the regime where no stochastic P-bifurcation takes place ($\lambda < -0.33$), although much less pronounced, see Figs. 2.3 and 2.8. In this regime, the probability distribution is smeared out, so no bimodal shape is observed.

For the explanation we investigate the part of the probability distribution, which transcends the radius where the saddle-node bifurcation of limit cycles takes place in the deterministic case ($D = 0$), see Fig. 2.1.

The noise can somehow visualize the ghost of the deterministic limit cycle before this saddle-node bifurcation takes place. The quantity

$$g(D) = \int_{r_0}^{\infty} P(r)dr \qquad (2.52)$$

measures that part of the distribution (Eq. (2.30)), which exceeds the critical radius $r_0 = \sqrt{\frac{1}{2}}$, where the deterministically stable limit cycle is born; therefore, we call $g(D)$ the ghost weight. Note that the integral (Eq. (2.52)) has to be evaluated numerically. Therefore, we use 100 values for the noise intensity D from the intervall $D \in [0.001, 0.1]$ with the stepsize $\Delta D = 0.001$ and estimate the ghost weight. It increases monotonically as a function of the noise, see Fig. 2.9. To see how this quantity changes due to the noise intensity, we calculate the corresponding derivative $\frac{dg(D)}{dD}$. Then the derivative is plotted for different λ, in the regime of the stochastic P-bifurcation and below this region, see Fig. 2.10.

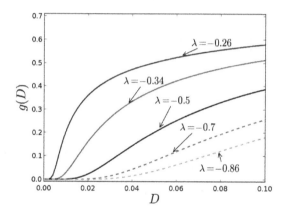

Figure 2.9: Ghost weight $g(D)$ (Eq. (2.52)) for different values of λ, calculated from the corresponding probability distribution (Eq. (2.30)).

In Fig. 2.10 we observe that the sharpest change of the ghost weight takes place in the regime of the stochastic bifurcation. But a remarkably high change can also be seen outside of this region. For this effect the interplay between the focus and the ghost of the stable limit cycle becomes important. With increasing noise intensity, the system can be driven further away from the deterministically stable focus located at the origin. Therefore, it will reach the critical radius r_0 at a certain noise strength; this can be visualised by the distribution. The presence of the ghost of the stable limit cycle gives rise to a second preferred region in the phase space, so the ghost weight rapidly increases at this noise strength, which results in a pronounced peak in the derivative with respect to the noise intensity. Figure 2.10 also shows that the most pronounced resonance-like behaviour occurs

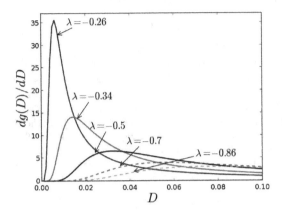

Figure 2.10: Derivative of the ghost weight $g(D)$ (Eq. (2.52)) for different λ, calculated from the corresponding probability distribution (Eq. (2.30)).

close to the saddle-node bifurcation in the parameter regime where the stochastic P-bifurcation takes place. For higher noise intensities, the change of the ghost weight is very small, because the main part of the distribution has already passed the critical radius.

Going further away from the saddle-node bifurcation of limit cycles, the resonance-like peak in the derivative of the ghost weight decreases and is shifted to higher noise intensities. The impact of the deterministically stable focus is getting stronger by increasing λ to higher negative values, so more noise is needed to drive the system to higher amplitudes. The attraction of the ghost of the limit cycle also decreases by moving away from the bifurcation, therefore, the change of the ghost weight is less pronounced. But it is still visible and shows the impact of the ghost, which results also in a weak maximum for the correlation time t_{cor} for values far away from the saddle-node bifurcation and in the absence of the stochastic P-bifurcation e.g. $\lambda = -0.5$, see Fig. 2.8.

In order to show that the ghost weight is not a unique measure of coherence, we also calculated the ghost weight and the corresponding derivative for the super-critical case according to the probability distribution given by Eq. (2.23). The derivative of the ghost weight also exhibits a clear visible maximum close to the supercritical Hopf bifurcation (Fig. 2.12), but we cannot observe coherence resonance. The maximum of the derivative of the ghost weight is much smaller than for the subcritical case (Fig. 2.14), but still visible. This is why we have to be careful when using this measure. This can somehow be compared with the signal-to-noise ratio in [7], where this measure shows a maximum also for the supercritical case. But this result is caused by the competition of two monotonic increasing functions and cannot be related to coherence resonance.

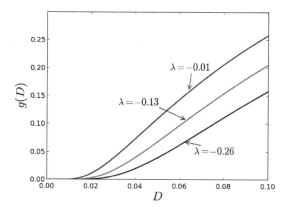

Figure 2.11: Ghost weight (Eq. (2.52)) in the supercritical case for different values of λ, calculated from the corresponding probability distribution (Eqs. (2.23, 2.24)).

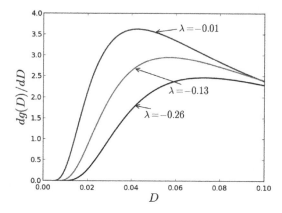

Figure 2.12: Derivative of the ghost weight (Eq. (2.52)) in the supercritical case for different values of λ, calculated from the corresponding probability distribution (Eqs. (2.23, 2.24)).

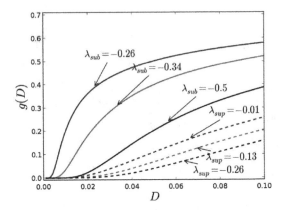

Figure 2.13: Ghost weight (Eq. (2.52)) for the subcritical (solid lines, $a = -1, b = 1$) and supercritical (dashed lines, $a = 1, b = 0$) case for different values of λ. The corresponding probability distributions are represented by Eq. (2.30) in the subcritical, and Eqs. (2.23, 2.24) in the supercritical case.

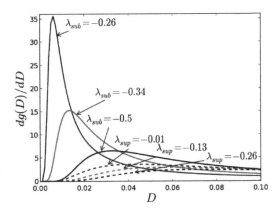

Figure 2.14: Derivative of the ghost weight (Eq. (2.52)) for the subcritical (solid lines, $a = -1, b = 1$) and supercritical (dashed lines, $a = 1, b = 0$) case for different values of λ. The corresponding probability distributions are represented by Eq. (2.30) in the subcritical, and Eqs. (2.23, 2.24) in the supercritical case.

Therefore, it is very important to mention the fact that the ghost weight is not a unique measure to detect coherence resonance. It will just be used for providing an explanation of the mechanism of coherence resonance in non-excitable systems in the absence of a stochastic P-bifurcation.

Chapter 3

Time-delayed feedback in nonlinear stochastic systems

In this chapter [1], we will discuss the stochastic Hopf normal form with time-delayed feedback. We want to study the influence of the time delay upon stochastic P-bifurcations and coherence resonance. First, as in chapter 2, a bifurcation and stability analysis is performed for the deterministic system.

3.1 Deterministic dynamics

The subcritical Hopf normal form with time-delayed feedback reads

$$\dot{z}(t) = (\lambda + i\omega_0 + s|z(t)|^2 - |z(t)|^4)z(t) - K(z(t) - z(t - \tau)). \tag{3.1}$$

λ is the deterministic bifurcation parameter, ω_0 is the intrinsic frequency and $z(t) \in \mathbb{C}$. K is the real-valued, positive coupling strength, τ describes the time delay, which will be used in units of the intrinsic time scale $T = \frac{2\pi}{\omega_0}$. We set the real-valued coefficient $s > 0$, which means that we restrict our investigations to the subcritical case. As mentioned in [7] and as we showed in chapter 2, coherence resonance only occurs in the subcritical Hopf normal form. The time-delayed coupling term is of the form of Pyragas control [10]. We will follow the methods used in [15, 16, 63] for the bifurcation and stability analysis. For the fixed point $z^*(t) = 0$ of Eq. (3.1), we find after linearisation (compare section (2.1), $z(t) \propto e^{\Lambda t}$) the characteristic equation

$$\Lambda = \lambda + i\omega_0 - K(1 - \exp(-\Lambda\tau)). \tag{3.2}$$

The subcritical Hopf bifurcation takes place at purely imaginary eigenvalues $\Lambda = i\Omega$. We decompose the equation above into real and imaginary part and obtain

---

[1]The main part of this chapter was published as [63] and is reused with kind permission of The European Physics Journal (EPJ) and Springer Science+Business Media.

the following conditions for the Hopf bifurcation

$$\Omega = \omega_0 - K\sin(\Omega\tau), \tag{3.3}$$

$$\lambda = K(1 - \cos(\Omega\tau)). \tag{3.4}$$

We observe that the Hopf bifurcation point is shifted to higher values of λ by increasing τ up to a half integer values of the intrinsic time scale; by further increasing, the bifurcation point moves back and reaches the original position at an integer value. To visualise this effect, we compute an expression for the representation in the (λ, τ)-plane.

Solving Eq. (3.3) for $\Omega\tau$ yields

$$\Omega\tau = \pm\arccos\left(\frac{K - \lambda}{K}\right) + 2\pi m \tag{3.5}$$

with $m \in \mathbb{Z}$. Using the trigonometric identity

$$\cos^2(\Omega\tau) + \sin^2(\Omega\tau) = 1, \tag{3.6}$$

we obtain from Eqs. (3.3, 3.4)

$$\left(\frac{K - \lambda}{K}\right)^2 + \left(\frac{\omega_0 - \Omega}{K}\right)^2 = 1. \tag{3.7}$$

The solution for Ω reads

$$\Omega = \omega_0 \mp \sqrt{K^2 - (K - \lambda)^2}. \tag{3.8}$$

Combining Eqs. (3.5, 3.8), eliminating Ω and solving for τ yields the following expression for the Hopf bifurcation lines

$$\tau_h = \frac{\pm\arccos\left(\frac{K-\lambda}{K}\right) + 2\pi m}{\omega_0 \mp \sqrt{K^2 - (K - \lambda)^2}}. \tag{3.9}$$

Next, we perform the bifurcation analysis for periodic states of rotating wave form

$$z(t) = r\exp(i\omega t), \tag{3.10}$$

where r denotes the positive amplitude and ω is the frequency. Inserting this ansatz into Eq. (3.1), we obtain

$$-r^4 + sr^2 = -\lambda + K(1 - \cos(\omega\tau)), \tag{3.11}$$

$$\omega = \omega_0 - K\sin(\omega\tau). \tag{3.12}$$

From Eq. (3.11) we find the limit cycles

$$r_1^* = \sqrt{\frac{s + \sqrt{s^2 + 4(\lambda + K(\cos(\omega\tau) - 1))}}{2}}, \tag{3.13}$$

$$r_2^* = \sqrt{\frac{s - \sqrt{s^2 + 4(\lambda + K(\cos(\omega\tau) - 1))}}{2}}. \tag{3.14}$$

To investigate the influence of the delay to the saddle-node bifurcation of limit cycles we rearrange Eq. (3.11) to

$$\left(r^2 - \frac{s}{2}\right)^2 = \lambda + \frac{s^2}{4} - K(1 - \cos(\omega\tau)) \tag{3.15}$$

and can observe that the right hand side of Eq. (3.15) has to be non-negative. So the saddle-node bifurcation of limit cycles takes place under the condition

$$\lambda + \frac{s^2}{4} - K(1 - \cos(\omega\tau)) = 0. \tag{3.16}$$

At this bifurcation point, the amplitude of the periodic state reads

$$r = \sqrt{\frac{s}{2}}. \tag{3.17}$$

In the same way as for the fixed point $z^*(t) = 0$ we obtain from Eq. (3.12) and Eq. (3.16) the saddle-node bifurcation lines

$$\tau_{sn} = \frac{\pm \arccos\left(\frac{K - \lambda - \frac{s^2}{4}}{K}\right) + 2\pi m}{\omega_0 \mp \frac{1}{4}\sqrt{(8K - 4\lambda - s^2)(4\lambda + s^2)}}. \tag{3.18}$$

The Hopf (Eq. (3.9)) and the saddle-node bifurcation lines (Eq. (3.18)) are presented in Fig. 3.1.

We observe that the positions, where the bifurcations take place, are shifted to higher values of the bifurcation parameter λ by increasing the delay time τ. The highest possible value of λ can be reached by an half integer delay time, whereas for integer delays the original bifurcation scenario is achieved.

For higher delay times ($\tau > 2$) the bifurcation scenario becomes more complicated. The character of the Hopf bifurcation can change from subcritical to supercritical. This aspect will be important in section 3.2.1. In the saddle-node bifurcation of limit cycles two periodic solutions exist with equal frequency given by Eq. (3.12). For higher delay times, more solutions arise from this equation, so we obtain at these saddle-node bifurcations two periodic states with different frequencies. The latter type of saddle-node bifurcation is named non-rigid bifurcation, whereas the other one is called rigid bifurcation, see [63] for a detailed study. Furthermore, the Hopf and saddle-node bifurcation lines can cross.

Now we want to investigate the stability of the fixed point. Equation (3.2) is a

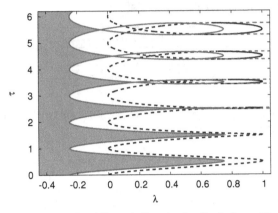

Figure 3.1: Hopf and saddle-node lines in the (λ, τ)-plane. Black: Hopf bifurcation lines (dashed: subcritical bifurcation, solid: supercritical bifurcation). Grey: saddle-node bifurcation lines (dashed: non-rigid bifurcation, solid: rigid bifurcation, see [63] for a detailed study). The parameter regime of interest, which is below the saddle-node bifurcation of the periodic solutions, is slightly shaded. Parameters: $K = 0.5$, $\omega_0 = 2\pi$, $s = 1$, for $m = 0, 1, .., 6$. Reused from [63] with kind permission of The European Physics Journal (EPJ).

transcendental equation for Λ and has an infinite number of solutions. For an analytical expression we can write

$$\Lambda_l = \lambda + i\omega_0 - K + \frac{1}{\tau} W_l \left[K\tau \exp(-(\lambda + i\omega_0 - K)\tau) \right] \qquad (3.19)$$

where W_l denotes the l-branch of the Lambert W-function [75]. The fixed point $z^*(t) = 0$ is unstable, when the largest real part of Λ_l becomes positive. Figure 3.2 displays the non-monotonic behaviour of the real part of Λ_l the as a function of the delay time τ: for small τ the main branch ($l = 0$) dominates the dynamics; further increasing of τ leads to more influence from the other branches. At half integer delays the focus is more stable, because the delay shifts the bifurcation scenario further away to higher values of λ. When choosing integer delays the focus becomes less stable in comparison to the half integer values and also to smaller integer values. This will become important for the investigation of coherence resonance, see section (3.2.3).
Next, we discuss the stability of the periodic states by linearising around the orbit:

$$z(t) = (r + \delta r) \exp(i\omega t + i\delta\phi). \qquad (3.20)$$

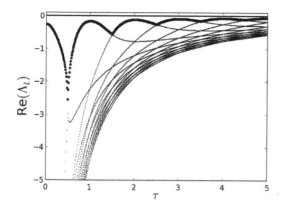

Figure 3.2: Real part of Λ_l according to Eq. (3.19). The main branch of the Lambert W-function ($l = 0$) is presented by the big dots. Only the branches for $l = -7, .., 7$ are shown here. Parameters: $\lambda = -0.26$, $K = 0.5$, $\omega_0 = 2\pi$.

Decomposing Eq. (3.1) in polar coordinates, the linearised system reads

$$\frac{d}{dt}\begin{pmatrix} \delta r(t) \\ \delta\phi(t) \end{pmatrix} = \begin{pmatrix} \lambda - 5r^4 + 3sr^2 - K & -Kr\sin(\omega\tau) \\ \frac{K}{r}\sin(\omega\tau) & -K\cos(\omega\tau) \end{pmatrix}\begin{pmatrix} \delta r(t) \\ \delta\phi(t) \end{pmatrix}$$
$$+ \begin{pmatrix} K\cos(\omega\tau) & Kr\sin(\omega\tau) \\ -\frac{K}{r}\sin(\omega\tau) & K\cos(\omega\tau) \end{pmatrix}\begin{pmatrix} \delta r(t-\tau) \\ \delta\phi(t-\tau) \end{pmatrix}. \qquad (3.21)$$

As in chapter 2, we make an exponential ansatz for the small deviations (δr, $\delta\phi \propto \exp(\Lambda t)$). So we obtain the following eigenvalue problem for Λ

$$\left| \begin{pmatrix} A & B \\ C & D \end{pmatrix} \right| = 0 \qquad (3.22)$$

with

$$A = -5r^4 + 3sr^2 - K(1 - \cos(\omega\tau)\exp(-\Lambda\tau)) + \lambda - \Lambda,$$
$$B = -Kr\sin(\omega\tau)(1 - \exp(-\Lambda\tau)),$$
$$C = \frac{K}{r}\sin(\omega\tau)(1 - \exp(-\Lambda\tau)),$$
$$D = -K\cos(\omega\tau)(1 - \exp(-\Lambda\tau)) - \Lambda. \qquad (3.23)$$

Now we have to insert the limit cycles from the Eqs. (3.13, 3.14) and to calculate the largest Floquet exponent. We have to do the calculation numerically because the characteristic equation is of transcendental type.

It turns out that the limit cycle corresponding to Eq. (3.13) is stable, the largest Floquet exponent is negative for different choices of τ and K, whereas the limit cycle corresponding to Eq. (3.14) is unstable, because the largest Floquet exponent is always positive.

It appears that this type of feedback term does not change the stability of the orbits, just the bifurcation scenario is shifted, see Fig. 3.1. For a change of the stability, a complex coupling parameter is needed, where a suitable choice of the coupling phase can lead to the stabilisation of an unstable orbit [16, 21].

3.2   Time-delayed feedback and noise

As shown in the previous section, the dynamical behaviour of nonlinear systems with time delayed feedback becomes more complex. Nevertheless, time-delayed feedback can be used to control the noise-induced properties of a system by a suitable choice of the time delay. This has been investigated in excitable neuronal systems with Gaussian white noise [25, 26] or correlated noise [76], non-excitable systems [52, 77], lasers [78, 79] or semiconductor structures [80–82]. The impact of a nonlinear coupling term was considered in [83].

Here, we want to investigate the modulation of coherence resonance in non-excitable systems, but we use different methods to achieve an analytical treatment as it was done in [84]. Our full stochastic delay differential equation now reads

$$\dot{z}(t) = (\lambda + i\omega_0 + s|z(t)|^2 - |z(t)|^4)z(t) - K(z(t) - z(t - \tau)) + \sqrt{2D}\xi(t) \quad (3.24)$$

where $D \geq 0$ is the noise intensity and $\xi(t)$ describes Gaussian white noise.

Numerical simulations of Eq. (3.24) for different delay times show a modulation of the probability distribution and the correlation time, see Figs 3.3 and 3.4. The amplitude probability distribution is changed in such a way that the bimodality vanishes for half integer delays. For integer delay times the bimodality is changed slightly so that more noise is needed to reach the same shape as for the non-delayed case. The result for the correlation time is similar to previous studies on coherence resonance in excitable systems [25, 26]: for the half integer delay times, the correlation time is very small, the effect of coherence resonance is suppressed. For integer delays, the effect is enhanced, one can observe a higher correlation time and a small shift of the optimal noise intensity.

As for the non-delayed case, one wants to have some analytical expressions for the measures of coherence and the probability distribution. But Eq. (3.24) describes a non-Markov process caused by the delay. More precisely infinite dimensional function spaces related with delay differential equation [9] make a direct calculation of the probability distribution as in section 2.2 (see also Appendix A) very difficult. There have been many studies to obtain the probability distribution for systems with time delay analytically. For linear systems, it is even possible to derive an exact expression [85]. Many other approximations are developed as small delay approximations [86–88] or multivariate Fokker-Planck equations for data analysis of nonlinear systems [89], but are mainly performed for linear equations.

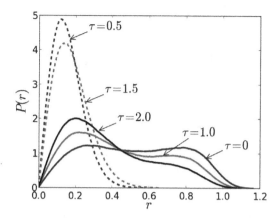

Figure 3.3: Numerical simulation of the amplitude probability distribution for different delay times, based on Eq. (3.24).
Parameters: $\lambda = -0.26$, $s = 1$, $\omega_0 = 2\pi$, $K = 0.5$, $D = 0.015$.

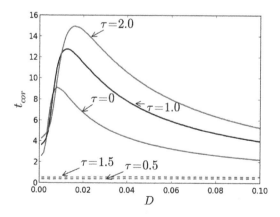

Figure 3.4: Numerical simulation of the correlation time for different delay times, based on Eq. (3.24).
Parameters: $\lambda = -0.26$, $s = 1$, $\omega_0 = 2\pi$, $K = 0.5$.

In the next section, the stochastic delay differential equation is reduced to an effective stochastic differential equation. The mathematical path for such a centre manifold reduction is shown in [90]. We will present a different way in terms of a multiple scale perturbation expansion [63, 91, 92].

3.2.1   Multiple scale perturbation expansion

We start with the assumption that the deterministic part of Eq. (3.24) for $D = 0$ undergoes a Hopf bifurcation

$$\lambda_0 = K(1 - \cos(\Omega\tau)), \quad \Omega = \omega_0 - K\sin(\Omega\tau). \tag{3.25}$$

The linear deterministic part of Eq. (3.24) describes harmonic oscillations of the form

$$z(t) = A\exp(i\Omega t). \tag{3.26}$$

We are interested in the dynamics close to the bifurcation ($\lambda = \lambda_0 + \delta\lambda$). Thus, a small expansion parameter ε is introduced and we make an amplitude modulated ansatz

$$z(t) = \varepsilon A(\varepsilon^4 t)\exp(i\Omega t). \tag{3.27}$$

All parts of Eq. (3.24) contribute equally to the stochastic dynamics. Therefore, we scale $\delta\lambda \to \varepsilon^4\delta\lambda$, $s \to \varepsilon^2 s$, and $D \to \varepsilon^6 D$. The noise is scaled as follows

$$\langle\xi(\varepsilon^4 t)\xi^*(\varepsilon^4 t')\rangle = 2\delta(\varepsilon^4(t - t'))$$
$$= \frac{2}{\varepsilon^4}\delta(t - t')$$
$$= \frac{1}{\varepsilon^4}\langle\xi(t)\xi^*(t')\rangle \tag{3.28}$$

$$\to \quad \xi(\varepsilon^4 t) = \frac{1}{\varepsilon^2}\xi(t). \tag{3.29}$$

The delay term is expanded in a Taylor series; it will turn out that even the delay contributes in the same way as the other parameters do:

$$A(\varepsilon^4(t - \tau)) = A(\varepsilon^4 t) - \varepsilon^4\tau A'(\varepsilon^4 t) + \mathcal{O}(\tau^2), \tag{3.30}$$

where $A'(\varepsilon^4 t)$ denotes the derivative due to the slow time scale $\varepsilon^4 t$.
Inserting all this into Eq. (3.24) we obtain

$$\varepsilon\exp(i\Omega t)(A(\theta)i\Omega + \varepsilon^4 A'(\theta))$$
$$= (\lambda_0 + \varepsilon^4\delta\lambda + i\omega_0 + s\varepsilon^2|\varepsilon A(\theta)|^2 - |\varepsilon A(\theta)|^4)\varepsilon A(\theta)\exp(i\Omega t)$$
$$- K\varepsilon\exp(i\Omega t)(A(\theta) - \exp(-i\Omega\tau)(A(\theta) - \varepsilon^4\tau A'(\theta))) + \sqrt{2D\varepsilon^6}\varepsilon^2\xi(\theta) \tag{3.31}$$

where $\theta = \varepsilon^4 t$ denotes the slow time scale. We obtain terms of order ε and of order ε^5 (leading order). The terms of order ε fulfil the conditions for the Hopf bifurcation (see Eq. (3.25)):

$$i\Omega = \lambda_0 + i\omega_0 - K(1 - (\cos(\Omega\tau) - i\sin(\Omega\tau))). \tag{3.32}$$

In leading order ε^5 we get

$$\exp(i\Omega t)A'(\theta) = (\delta\lambda + s|A(\theta)|^2 - |A(\theta)|^4)A(\theta)\exp(i\Omega t)$$
$$-K\exp(i\Omega t)\exp(-i\Omega\tau)\tau A'(\theta) + \sqrt{2D}\xi(\theta). \tag{3.33}$$

Rewriting Eq. (3.33) we obtain

$$A'(\theta) = \frac{(\delta\lambda + s|A(\theta)|^2 - |A(\theta)|^4)A(\theta)}{1 + K\tau\exp(-i\Omega\tau)} + \frac{\sqrt{2D}\zeta(\theta)}{1 + K\tau\exp(-i\Omega\tau)}, \tag{3.34}$$

where we use the assumption that the isotropic noise is not affected by a complex phase transformation $\zeta(\theta) = \xi(\theta)\exp(-i\Omega_0 t)$.
The full stochastic delay differential equation (Eq. (3.1)) is reduced to an effective stochastic differential equation (Eq. (3.34)). Therefore, we can derive the stationary amplitude probability distribution, because Eq. (3.34) describes a Markov process. To make the noise intensity real-valued, we take advantage of the fact that a complex phase can be absorbed by the isotropic noise: $1 + K\tau\exp(-i\Omega\tau) = |1 + K\tau\exp(-i\Omega\tau)|\exp(-i\alpha)$, $\tilde{\xi}(t) = \zeta(t)\exp(-i\alpha)$, so we find

$$A'(\theta) = \frac{1 + K\tau\exp(i\Omega\tau)}{|1 + K\tau\exp(-i\Omega\tau)|^2}\left(\delta\lambda + |A(\theta)|^2 - |A(\theta)|^4\right)A(\theta)$$
$$+ \sqrt{\frac{2D}{|1 + K\tau\exp(-i\Omega\tau)|^2}}\tilde{\xi}(\theta). \tag{3.35}$$

We transform to polar coordinates $A = r\exp(i\varphi)$, and derive the Fokker-Planck equation as already done in section (2.2) (see Appendix A). The stationary amplitude probability distribution reads

$$P(r) = Nr\exp\left(\frac{r^2}{D_{eff}}\left(\frac{\delta\lambda}{2} + \frac{sr^2}{4} - \frac{r^4}{6}\right)\right) \tag{3.36}$$

with

$$\delta\lambda = \lambda - K(1 - \cos(\Omega\tau)),$$
$$D_{eff} = \frac{D}{1 + K\tau\cos(\Omega\tau)}, \tag{3.37}$$

and N denotes the normalisation factor. We observe that the bifurcation parameter λ and even the noise intensity are modulated by the time delay.
To check the validity of this approximation, we compare the result with numerics for integer and half-integer delay times. Figure 3.5 shows the comparison between numerics and analytics for integer delay times and we observe that both are in good agreement. The half integer values are shown in Fig. 3.6: for $\tau = 0.5$ the curves display excellent agreement. For the other values there is a high remarkable

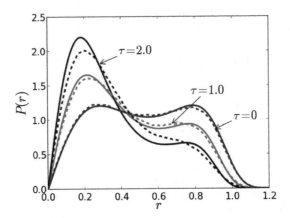

Figure 3.5: Amplitude probability distribution, calculated analytically (solid, Eqs. (3.36, 3.37)) and numerically (dashed) for different integer delay times. Parameters: $K = 0.5$, $\omega_0 = 2\pi$, $D = 0.015$, $\lambda = -0.26, s = 1$.

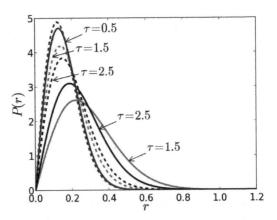

Figure 3.6: Amplitude probability distribution, calculated analytically (solid, Eqs. (3.36, 3.37)) and numerically (dashed) for different half integer delay times. Parameters: $K = 0.5$, $\omega_0 = 2\pi$, $D = 0.015$, $\lambda = -0.26, s = 1$.

deviation between numerics and analytics. Nevertheless, the suppression of the bimodality is captured by the analytical approximation. Although we made the approach with the assumption being close to the bifurcation, it displays the correct behaviour in a wide range of parameter values, at least qualitative, e.g. for $\tau = 1.5$. The approach is limited to small delay values, because we made use of a Taylor expansion of the delay-term (Eq. (3.30)). The scaling factor $1 + K\tau \cos(\Omega\tau)$ of the noise strength (Eq. (3.37)) plays also a crucial role. We have to take care, because this factor can change the sign; this happens when the product of coupling strength and delay time exceed the value one. Then we would obtain a negative effective noise strength and the bifurcation scenario would change from sub- to supercritical (see Fig. 3.1). As already mentioned in the deterministic bifurcation analysis, the transcendental equation for the frequency (Eq. (3.25)) provides more than one solution for higher values of the delay. Using such a solution $\Omega \neq 2\pi$, we obtain again a positive effective noise strength and a good result for our approximation, see Fig. 3.6 for the value $\tau = 2.5$.

To improve the approach, we should use higher order terms in the centre manifold reduction to describe the impact of the delay more appropriately or we have to use higher orders from the Taylor series (Eq. (3.30)).

Next, we calculate the stochastic bifurcation diagram from the result based on the analytical approximation (Eq. (3.36)). In Fig. 3.7, we observe that the delay shifts the "stochastic triangle" to higher values of λ up to the half integer delay time and back. The shape is also affected by increasing the delay, which results from the different solutions for Ω (Eq. (3.25)) and the scaled noise intensity (Eq. (3.37)). As we will see in section 3.2.3, coherence resonance is also influenced by these results: the most pronounced coherence resonance can be found in the regime of a bimodal shaped probability distribution (section 2.4.2 and [62]). If the time delay is increased, we have to adjust λ to stay inside of the triangle, see Fig. 3.7. This result of the shifted stochastic bifurcation diagram is not surprising, because the deterministic bifurcation scenario is also shifted in the same way, see Fig. 3.1.

In our work we only focussed on the case with additive noise; an investigation of multiplicative noise in stochastic delay differential equations and the analytical derivation of stationary probability distributions is shown in [93–95].

3.2.2 Power spectral density and correlation properties

As already done in section 2.1, we can linearise our system in the same way as in the non-delayed case [63, 71]. Again, we replace the nonlinear terms in Eq. (3.24) by an effective linear term (see Eqs. (2.37 - 2.39)).
Therefore, we obtain

$$\dot{z}(t) = (\tilde{\lambda} + i\omega_0)z(t) - K(z(t) - z(t - \tau)) + \sqrt{2D}\xi(t) \qquad (3.38)$$

with $\tilde{\lambda} = \lambda + \alpha$ and $\alpha = \frac{s\langle|z(t)|^4\rangle - \langle|z(t)|^6\rangle}{\langle|z(t)|^2\rangle}$.

The solution of a linear stochastic delay differential equation (Eq. (3.38)) can be expressed in closed form, see [85]. To calculate the measures of coherence, we adapt

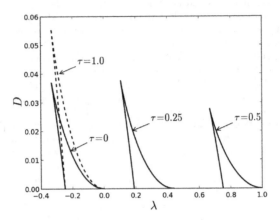

Figure 3.7: Stochastic bifurcation diagram for different time delays based on Eq. (3.36).
Parameters: $K = 0.5$, $\omega_0 = 2\pi$, $s = 1$.

the methods presented in [63, 90]. The characteristic equation corresponding to the deterministic part of Eq. (3.38) reads

$$\Lambda = (\tilde{\lambda} + i\omega_0) - K(1 - \exp(-\Lambda\tau)), \tag{3.39}$$

and can be solved by using the Lambert W-function (see Eqs.(3.2, 3.19))

$$\rightarrow \quad \Lambda_l = \frac{W_l(K\tau\exp(-(\tilde{\lambda} + i\omega_0 - K)\tau))}{\tau} + \tilde{\lambda} + i\omega - K. \tag{3.40}$$

The solution of Eq. (3.38) can be expressed in terms of eigenmodes $z(t) = \sum_l C_l(t)$, where the time evolution of the coefficients is determined by the linear stochastic differential equation

$$\dot{C}_l(t) = \Lambda_l C_l(t) + \frac{\sqrt{2D}\xi(t)}{N_l} \tag{3.41}$$

with the normalisation factor $N_l = 1 + K\tau\exp(-\Lambda_l\tau)$ (see Appendix C for a detailed derivation). We end up with the solution for a single mode

$$C_l(t) = C_0\exp(\Lambda_l(t)) + \frac{\sqrt{2D}}{N_l}\int_0^t \exp(\Lambda_l t'')\xi(t - t'')dt'' \tag{3.42}$$

which can be governed by using variation of constants. The stationary solution of Eq. (3.38) ($t \rightarrow \infty$) reads

$$z(t) = \sqrt{2D}\int_0^\infty T(t')\xi(t - t')dt', \tag{3.43}$$

where

$$T(t) = \sum_l \frac{\exp(\Lambda_l t)}{N_l}. \tag{3.44}$$

For the calculation of the autocorrelation function, we consider some relations between the complex variable and the real and imaginary part:

$$x(t) = \frac{z(t) + z^*(t)}{2}, \quad y(t) = \frac{z(t) - z^*(t)}{2i}. \tag{3.45}$$

Using the properties of the noise (Eq. (2.19)), we can write for the autocorrelation function of the real and the imaginary part

$$\langle x(t)x(0) \rangle = \langle y(t)y(0) \rangle = \frac{1}{2}\mathrm{Re}\langle z(t)z^*(0) \rangle. \tag{3.46}$$

Note that the variable t denotes here the shifting variable. We use the stationary expression of the autocorrelation function, so it depends just on the shifting argument. The brackets $\langle \ldots \rangle$ denote the ensemble average. We compute the autocorrelation function of the complex variable $z(t)$ as follows:

$$
\begin{aligned}
\langle z(t)z^*(0) \rangle &= 2D \sum_{l,l'} \int_0^\infty \int_0^\infty \frac{\exp(\Lambda_l t') \exp(\Lambda_{l'}^* t'')}{N_l N_{l'}^*} \underbrace{\langle \xi(t-t')\xi^*(-t'') \rangle}_{2\delta(t-t'+t'')} \, dt' dt'' \\
&= 4D \sum_{l,l'} \int_0^\infty \frac{\exp(\Lambda_l t) \exp((\Lambda_l + \Lambda_{l'}^*)t'')}{N_l N_{l'}^*} dt'' \\
&= 4D \sum_{l,l'} \frac{\exp(\Lambda_l t)}{N_l N_{l'}^*(-\Lambda_l - \Lambda_{l'}^*)}. \tag{3.47}
\end{aligned}
$$

From the Eqs. (3.46, 3.47) we get the autocorrelation function of the real part

$$\langle x(t)x(0) \rangle = 2D \sum_{l,l'} \mathrm{Re} \frac{\exp(\Lambda_l t)}{N_l N_{l'}^*(-\Lambda_l - \Lambda_{l'}^*)}. \tag{3.48}$$

Now we are able to investigate the correlation properties of our system, e.g. the power spectral density via the Wiener-Khinchin theorem and the correlation time. We start with the power spectral density, so the Fourier transform of the autocorrelation function (Eq. (3.48)) is estimated (detailed calculation in the Appendix D):

$$
\begin{aligned}
S_{xx}(\omega) &= \int_{-\infty}^\infty e^{i\omega t} \langle x(t)x(0) \rangle dt \\
&= 2D\mathrm{Re} \sum_{l,l'} \frac{-2\Lambda_l}{\Lambda_l^2 + \omega^2} \frac{1}{N_l N_{l'}^*(-\Lambda_l - \Lambda_{l'}^*)}. \tag{3.49}
\end{aligned}
$$

Using the Laplace transform, it is also possible to give a closed expression for the power spectral density, see [63].
The validity of Eq. (3.49) is tested numerically for different system parameters.

For different time delays one example and a comparison with numerics is shown in Fig. 3.8. The power spectral density shows the central peak at the resonance

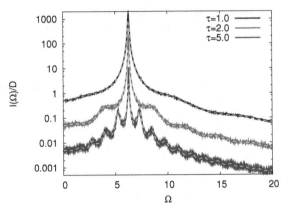

Figure 3.8: Normalised power spectral densities for different integer delay times. The dashed white lines correspond to Eq. (3.49) and the solid curves represent the result of the numerical simulation from Eq. (3.24). The top and bottom spectral density are shifted for a better visibility. Parameters: $\lambda = -0.26$, $K = 0.5$, $\omega_0 = 2\pi$, $D = 0.015$, $s = 1$. Reused from [63] with kind permission of The European Physics Journal (EPJ).

frequency $\omega = \omega_0$ and additional peaks below and above the central peak for increasing delay times. The approximate result from Eq. (3.49) shows excellent agreement with the numerics, even for high values as $\tau = 5$.

As already mentioned in the previous chapter, the power spectral density can also be calculated from a linear-response ansatz [74]. Other examples of computing the power spectral density regarding to a delay differential equation can be found in [52, 77].

To show the modulation of coherence resonance, we calculate the correlation time. Therefore, we use the autocorrelation function for the real part (Eq. (3.48)). As mentioned in section 2.4.1 (Eq. (2.35)), the correlation time can be defined as

$$t_{cor} = \int_0^\infty |\Psi_{xx}(t)| dt, \qquad (3.50)$$

where $\Psi_{xx}(t) = \frac{1}{\langle x(0)x(0) \rangle} \langle x(t)x(0) \rangle$ describes the normalised autocorrelation function. In the following calculation we use the abbreviation $M = \frac{1}{\langle x(0)x(0) \rangle}$. With Eq.

(3.48) we get

$$t_{cor} = \int_0^\infty |\Psi_{xx}(t)|dt$$

$$= |M| \int_0^\infty \left| \mathrm{Re} \sum_{ll'} \frac{\exp(\Lambda_l t)(N_l^* N_{l'}(-\Lambda_l^* - \Lambda_{l'}))}{\left|N_l N_{l'}^*(-\Lambda_l - \Lambda_{l'}^*)\right|^2} \right| dt$$

$$= |M| \int_0^\infty \left| \sum_{ll'} \frac{\exp(\gamma t)[A\cos(wt) + B\sin(wt)]}{A^2 + B^2} \right| dt, \qquad (3.51)$$

where we used in the last step

$$\mathrm{Re}\left[\exp(\Lambda_l t)(N_l^* N_{l'}(-\Lambda_l^* - \Lambda_{l'}))\right] = \exp(\gamma t)[\cos(wt)A + \sin(wt)B] \qquad (3.52)$$

with the abbreviations

$$\gamma = \mathrm{Re}(\Lambda_l), \ \ w = \mathrm{Im}(\Lambda_l), \ \ A = \mathrm{Re}(N_l^* N_{l'}(-\Lambda_l^* - \Lambda_{l'})),$$

$$\text{and } B = \mathrm{Im}(N_l^* N_{l'}(-\Lambda_l^* - \Lambda_{l'})).$$

To solve the integral (Eq. (3.51)), we use the triangle inequality $|\sum_k a_k| \leq \sum_k |a_k|$, and we obtain

$$|M| \int_0^\infty \left| \sum_{ll'} \frac{\exp(\gamma t)[A\cos(wt) + B\sin(wt)]}{A^2 + B^2} \right| dt$$

$$\leq \ |M| \int_0^\infty \sum_{ll'} \left| \frac{\exp(\gamma t)[A\cos(wt) + B\sin(wt)]}{A^2 + B^2} \right| dt$$

$$= \ |M| \sum_{ll'} \int_0^\infty \left| \frac{\exp(\gamma t)[A\cos(wt) + B\sin(wt)]}{A^2 + B^2} \right| dt$$

$$= \ |M| \sum_{ll'} \frac{1}{A^2 + B^2} \int_0^\infty \exp(\gamma t) |(A\cos(wt) + B\sin(wt))| dt. \qquad (3.53)$$

We rewrite the trigonometric functions of the integrand (Eq. 3.53) to obtain one cosine term, which can be replaced by a filling factor (see section 2.4) for an easier evaluation of the integral:

$$A\cos(\omega t) + B\sin(\omega t) = A\underbrace{\cos(\omega t)}_{\phi_1} + B\underbrace{\cos\left(\omega t - \frac{\pi}{2}\right)}_{\phi_2}$$

$$= \sqrt{A^2 + B^2 + 2AB\cos\left(\frac{\pi}{2}\right)}\cos(\omega t + \phi)$$

$$= \sqrt{A^2 + B^2}\cos(\omega t + \phi) \qquad (3.54)$$

with $\tan \phi = \frac{A \sin(\phi_1) + B \sin(\phi_2)}{A \cos(\phi_1) + B \cos(\phi_2)}$.

Therefore, we can solve the integral

$$\int_0^\infty \exp(\gamma t) |(A \cos(wt) + B \sin(wt))| \, dt$$

$$= \int_0^\infty \exp(\gamma t) \left| \sqrt{A^2 + B^2} \cos(wt + \phi) \right| dt$$

$$= \int_0^\infty \exp(\gamma t) \sqrt{A^2 + B^2} \left| \cos(wt + \phi) \right| dt$$

$$\approx \int_0^\infty \exp(\gamma t) \sqrt{A^2 + B^2} \frac{2}{\pi} dt$$

$$= \sqrt{A^2 + B^2} \frac{2}{\pi} \frac{-1}{\gamma}. \tag{3.55}$$

The final result is

$$t_{cor} = \frac{2}{\pi} |M| \sum_{ll'} \frac{\sqrt{A^2 + B^2}}{A^2 + B^2} \frac{-1}{\gamma}$$

$$= Z \sum_{ll'} \frac{1}{\left| N_l N_{l'}^*(-\Lambda_l - \Lambda_{l'}^*) \right| \text{Re}(-\Lambda_l)} \tag{3.56}$$

with $Z = \frac{2}{\pi} |M|$. The comparison between numerics and the analytical expression from Eq. (3.56) is shown in Fig. 3.9.

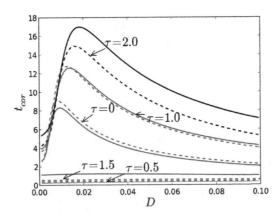

Figure 3.9: Correlation time for different time delays τ. The solid line corresponds to Eq. (3.56), but only the main branch ($l = 0$) of the Lambert W-function is used; the dashed lines denotes the numerics. Parameters: $K = 0.5$, $\lambda = -0.26$, $\omega_0 = 2\pi$, $s = 1$.

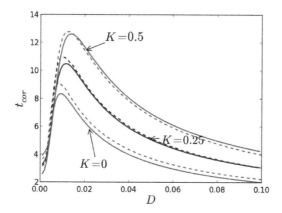

Figure 3.10: Correlation time for different values of K. The solid line corresponds to Eq. (3.56), but only the main branch ($l = 0$) of the Lambert W-function is used; the dashed lines denotes the numerics.
Parameters: $\tau = 1.0$, $\lambda = -0.26$, $\omega_0 = 2\pi$, $s = 1$.

Equation (3.56) captures well the behaviour of the correlation time for different time delays: as already mentioned for Fig. 3.4, the correlation time is enhanced for integer delay times and decreases for half integer delays. Here, we only plot the main branch ($l = 0$) of the Lambert W-function, because it dominates the dynamics for small time delays (see Fig. 3.2). For $\tau = 1.5$ and $\tau = 2.0$ a change in the analytical result can be observed by adding more branches but the differences are not that significant.

We also calculate the correlation time for different coupling strengths and a fixed time delay. Figure 3.10 displays that coherence resonance is enhanced by a higher coupling strength and also the optimal noise intensity is shifted to higher noise strengths.

3.2.3   Mechanism of coherence resonance in delayed systems

Now we want to discuss the modulation of the correlation time caused by the time delay.

We observe in Figs. 3.4 and 3.9 that the correlation time t_{cor} exhibits a pronounced maximum at a certain optimal noise intensity only for integer delays. For half integer delays this maximum is suppressed. Furthermore, the optimal noise intensity D_{opt} is shifted to higher values of D for larger time delays.

The aspect of the enhancement and suppression of coherence resonance can be related to the stability of the deterministic focus. We consider a system that is able to show coherence resonance. In [25] it is pointed out that noise-induced oscillations are more regular, the less stable the eigenmode is. In Fig. 3.2 it is shown that the focus is more stable for half integer delays and thus, it is more difficult to excite the system. Therefore, coherence resonance is suppressed. For the integer delay times, the focus becomes less stable so that it is easier to perturb the system, which means that the trajectory is kicked out of the stable focus. So coherence resonance is enhanced. As already mentioned, this kind of modulation could also be observed for excitable systems [25, 26].

For the explanation of the shift of the optimal noise intensity, we use the probability distribution (Eq. (3.36)) and plot the bifurcation diagram for the integer delays to compare it with the non-delayed case.

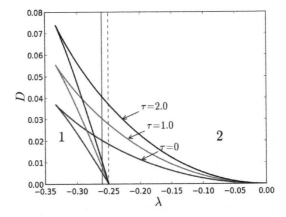

Figure 3.11: Stochastic bifurcation diagram for different integer time delays. The dashed line splits the figure up into the two deterministic regimes, compare Fig. 2.3. The black solid line shows the fixed λ value, where the noise intensity is increased (compare Fig. 2.4).
Parameters: $K = 0.5$, $\omega_0 = 2\pi$, $s = 1$.

Figure 3.11 shows that the lower bifurcation line is steeper for higher delays. For a fixed λ a higher value of the noise strength D is needed to reach the regime of bimodality. This is also connected with the scaling of the noise intensity by the delay (see Eq. (3.37)): to obtain the same value of the effective noise intensity D_{eff} for $\tau = 0$ and $\tau = 2$, the value D has to be higher for the case $\tau = 2$. As we mentioned in section 2.4, coherence resonance occurs in a most pronounced way in the regime of bimodality [62]. Therefore, the optimal noise intensity is shifted to higher values of D.

To complete the discussion about the mechanism of coherence resonance, we also

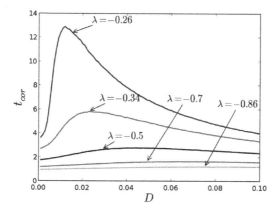

Figure 3.12: Correlation time for different values of λ, numerically calculated (Eq. (3.24)).
Parameters: $\tau = 1.0$, $K = 0.5$, $\omega_0 = 2\pi$, $s = 1$.

investigate the correlation time for parameter values outside the region of bimodality. The result is given in Fig. 3.12 and it displays the same behaviour as already shown in chapter 2, see Fig. 2.8: coherence resonance is most pronounced in the regime of bimodality, but can still be observed outside of this region.

To answer the question why coherence resonance can still occur outside of the regime of a bimodal probability distribution, we calculate again the ghost weight (Eq. (2.52)) due to the approximate probability distribution (Eq. (3.36)). The ghost weight was introduced in section 2.4.2 to measure the part of the distribution, which exceeds the critical radius, where the saddle-node bifurcation of limit cycles takes place for $D = 0$. The corresponding derivative displays how this part of the distribution changes by increasing the noise intensity.

The results of the investigations, including the time-delay, are given in the Figs. 3.13 and 3.14. We observe the same result as in the non-delayed case: the derivative of the ghost weight exhibits the resonance-like behaviour, which decreases for more negative values of λ and is most pronounced close to the saddle-node bifurcation of limit cycles.

Next, we compare the derivative of the ghost weight for different values of the time delay; this is shown in Fig. 3.15. Comparing the integer values of τ, we can observe that the resonance-like peak decreases for higher values of τ.

A similar result is obtained for different coupling strengths, which can also be connected to the effective noise strength (Eq. (3.57)), see Fig. 3.16; for a smaller coupling strength K the derivative of the ghost weight is more pronounced.

The decrease of the maximum of the ghost weight derivative due to higher time delays and higher coupling strengths is connected with the rescaling of the noise

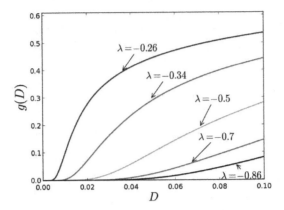

Figure 3.13: Ghost weight for different values of the bifurcation parameter λ, calculated from the approximate probability distribution (Eq. (3.36)).
Parameters: $\tau = 1.0$, $K = 0.5$, $\omega_0 = 2\pi$, $s = 1$.

intensity (Eq. (3.37))

$$D_{eff} = \frac{D}{1 + K\tau\cos(\Omega\tau)}. \tag{3.57}$$

In all numerical simulations and analytical plots we increased the parameter D
with the step size $\Delta D = 0.001$. For the non-delayed case ($K = 0$, $\tau = 0$) the
parameter D_{eff} rises with the same step size $\Delta D_{eff} = 0.001$, see Eq. (3.57).
Setting $K = 0.5$ and $\tau = 1.0$ we get from Eq. (3.57) (Ω is calculated via the
transcendental equation (3.25))

$$D_{eff} = \frac{2D}{3}. \tag{3.58}$$

Again by increasing D with the step size $\Delta D = 0.001$ we obtain the effective step
size $\Delta D_{eff} = 0.00067$. So the increase of the effective noise intensity is slower
compared to the non-delayed case. By choosing a higher value for the time delay
$\tau = 2.0$ and $K = 0.5$ the effective step size reads $\Delta D_{eff} = 0.0005$, which describes
again a slower increase of the effective noise intensity.
For the half integer delays it is the other way around: we have $\Delta D_{eff} = 0.0013$ for
$\tau = 0.5$ and $\Delta D_{eff} = 0.004$ for $\tau = 1.5$. This is why we also observe a maximum
for the case of nearly suppressed coherence resonance for $\tau = 1.5$, although we are
far away from the saddle-node bifurcation of limit cycles.
The various increase of the part of the probability distribution exceeding the critical radius is shown for different integer time delays in Figs. 3.17 - 3.19: we increase
the noise intensity from $D = 0.005$ to $D = 0.01$ with a step size of $\Delta D = 0.001$ and
can observe that the part of the distribution exceeding the critical radius $r_0 = \sqrt{\frac{1}{2}}$

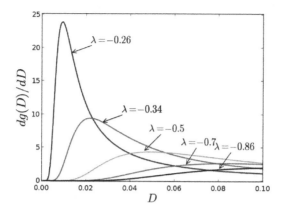

Figure 3.14: Derivative of the ghost weight for different values of the
bifurcation parameter λ, calculated from the approximate
probability distribution (Eq. (3.36)).
Parameters $\tau = 1.0$, $K = 0.5$, $\omega_0 = 2\pi$, $s = 1$.

shows the strongest increasing for the non-delayed case and the slowest increase is
achieved for the highest time delay, here $\tau = 2$.
The effects shown in the Figs. 3.15 and 3.16 reflect the opposite behaviour compared to the correlation time by varying the delay time or the coupling strength,
see Figs. 3.9 and 3.10. A comparison of the derivative of the ghost weight between
different delay times τ or different coupling strength K can lead to confusion.
Therefore, we have to point out that this measure cannot be used to conclusively
determine higher or lower regularity between two different delay times or coupling
strengths.

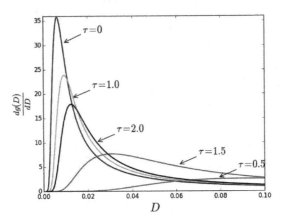

Figure 3.15: Derivative of the ghost weight for different values of the time delay τ, calculated from the approximate probability distribution (Eq. (3.36)).
Parameters: $\lambda = -0.26$, $K = 0.5$, $\omega_0 = 2\pi$, $s = 1$.

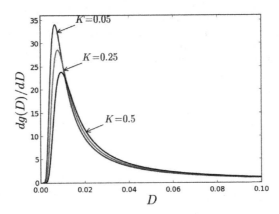

Figure 3.16: Derivative of the ghost weight for different values of the coupling strength K, calculated from the approximate probability distribution (Eq. (3.36)).
Parameters: $\lambda = -0.26$, $\tau = 1.0$, $\omega_0 = 2\pi$ $s = 1$.

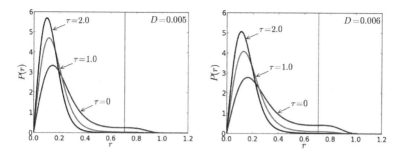

Figure 3.17: Amplitude probability distribution (Eq. (3.36)). The vertical line denotes the critical radius $r_0 = \sqrt{1/2}$. Parameters: $\lambda = -0.26$, $\omega_0 = 2\pi$, $s = 1$, and for $\tau \neq 0$ $K = 0.5$.

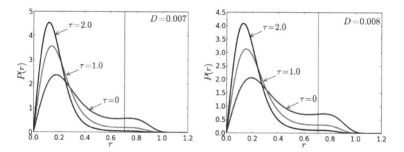

Figure 3.18: Amplitude probability distribution (Eq. (3.36)). The vertical line denotes the critical radius $r_0 = \sqrt{1/2}$. Parameters: $\lambda = -0.26$, $\omega_0 = 2\pi$, $s = 1$, and for $\tau \neq 0$ $K = 0.5$.

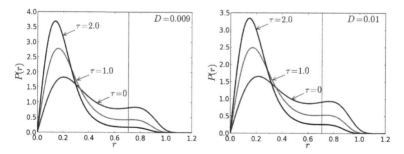

Figure 3.19: Amplitude probability distribution (Eq. (3.36)). The vertical line denotes the critical radius $r_0 = \sqrt{1/2}$. Parameters: $\lambda = -0.26$, $\omega_0 = 2\pi$, $s = 1$, and for $\tau \neq 0$ $K = 0.5$.

Chapter 4

Coupled stochastic systems

In the previous chapters, we investigated the modulation of coherence resonance and stochastic P-bifurcations by time-delayed feedback. For these studies, we made use of suitable methods such as statistical linearisation and a multiple scaling technique.

The aim of this chapter is to study the possible modulation of the noise effects in delayed coupled oscillator systems. For this purpose, we will try to extent our methods to coupled stochastic delay differential equations.

4.1 Stochastic bifurcations in coupled oscillators

A system of N stochastic oscillators with delayed coupling can be written in the following form

$$\dot{z}_k(t) = f(z_k(t)) + \sqrt{2D}\xi_k(t) - K\sum_{j=1}^{N} G_{kj}(z_k(t) - z_j(t - \tau)), \qquad (4.1)$$

where k is an index denoting the k-th oscillator and $f(z_k(t))$ is a function describing the local deterministic dynamics of this node, which is a nonlinear function in our case for the Hopf normal form $f(z_k(t)) = (\lambda + i\omega_0 + s|z_k(t)|^2 - |z_k(t)|^4)z_k(t)$. λ denotes the deterministic bifurcation parameter, ω_0 the intrinsic frequency, $s > 0$ is the cubic coefficient and we just investigate the subcritical case. D describes the noise intensity, $\xi_k(t)$ is independent Gaussian white noise with the property $\langle \xi_k(t)\xi_l^*(t') \rangle = \delta_{kl}\delta(t - t')$, τ is the delay time, and K is the real-valued coupling constant.

As already mentioned in section 3.1, a complex coupling strength $\sigma = Ke^{i\beta}$, where K denotes the amplitude and β the phase, is useful in the context of stabilisation. [37] and [96] are nice examples in this context with investigations of coupled Hopf normal forms.

We concentrate on the case $N = 2$ for Eq. (4.1)

$$\dot{z}_1(t) = (\lambda + i\omega_0 + s|z_1(t)|^2 - |z_1(t)|^4)z_1(t) - K(z_1(t) - z_2(t - \tau)) + \sqrt{2D}\xi_1(t), \quad (4.2)$$

$$\dot{z}_2(t) = (\lambda + i\omega_0 + s|z_2(t)|^2 - |z_2(t)|^4)z_2(t) - K(z_2(t) - z_1(t - \tau)) + \sqrt{2D}\xi_2(t). \quad (4.3)$$

It is a very difficult task to solve the coupled Eqs. (4.2, 4.3) for the oscillators z_1 and z_2. Therefore, we have to find another way to obtain the results for the stochastic dynamics of the coupled system.

An expansion in network modes was performed in [97] and the resulting equations were solved. This was done for damped oscillators and Hopf normal forms; the latter were linearised by a mean-field approximation. Because of the linear equations for the local dynamics, the decoupling of the system could be done in a straightforward way by diagonalising the coupling matrix.

In the previous chapters, we were also able to linearise our equation, even for the delayed case. But for this purpose we used the corresponding probability distribution. Here, a direct derivation of the probability distribution is again very difficult. So we are not able to decouple our coupled oscillator system by a direct calculation.

We make use of the fact that mutually delay-coupled Hopf normal forms show different types of network patterns: there is an in-phase solution and out-of-phase motions, which can be described as anti-phase or splay-state solutions. It is possible to control which of the network motions is performed by the choice of the time delay.

It is important to know that for single systems with time-delayed feedback [98] but also for networks with delayed-coupling [99] it was shown that the periodic solutions form branches due to the time delay parameter τ. An increase of τ can lead to the coexistence of multiple stable and unstable branches.

The two Eqs. (4.2, 4.3) exhibit in the deterministic case ($D = 0$) an in-phase and an anti-phase solution depending on the value of the time delay. We assume deterministic solutions of rotating wave form

$$z_k(t) = re^{i\omega t}, \quad (4.4)$$

where r denotes the positive amplitude and ω the frequency. This ansatz was also successfully used for coupled stochastic Hopf normal forms in [74]. For integer values of τ, we will get the in-phase solution, whereas for half integer values of τ the anti-phase solution describes the dynamics of the coupled system. Applying the in-phase solution ansatz $z_S = z_1 = z_2$ to the Eqs. (4.2, 4.3), we obtain two equations of the form

$$\dot{z}_S(t) = (\lambda + i\omega_0 + s|z_S(t)|^2 - |z_S(t)|^4)z_S(t) - K(z_S(t) - z_S(t - \tau)) + \sqrt{2D}\xi_S(t). \quad (4.5)$$

Note that we only adapted an deterministic ansatz to the stochastic equations and ignored for now the stochastic input coming from the coupled oscillator. We expect that this approach will work especially in the low noise limit ($D \to 0$).

We investigated such a type of stochastic delay differential equation already in chapter 3 and performed also the deterministic bifurcation analysis (see Fig. 3.1).

Therefore, the stationary amplitude probability distribution reads

$$P(r) = Nr \exp\left(\frac{r^2}{D_{eff}}\left(\frac{\delta\lambda}{2} + \frac{sr^2}{4} - \frac{r^4}{6}\right)\right), \tag{4.6}$$

with

$$\delta\lambda = \lambda - K(1 - \cos(\Omega\tau)), \quad D_{eff} = \frac{D}{1 + K\tau\cos(\Omega\tau)}. \tag{4.7}$$

For half integer τ we obtain the anti-phase solution $z_A = z_1 = -z_2$, which reads

$$\dot{z}_A(t) = (\lambda + i\omega_0 + s|z_A(t)|^2 - |z_A(t)|^4)z_A(t) - K(z_A(t) + z_A(t-\tau)) + \sqrt{2D}\xi_A(t). \tag{4.8}$$

This equation has a slightly different coupling term, but we can calculate the corresponding stationary probability distribution for the amplitude by using the methods from section 3.2. We just point out the main steps of the calculation: the conditions for the Hopf bifurcation are

$$\lambda = K(1 + \cos(\Omega\tau)), \quad \Omega = \omega + K\sin(\Omega\tau). \tag{4.9}$$

Using the multiple scaling technique, we derive the corresponding effective equation of motion

$$A'(\theta) = \frac{(\delta\lambda + s|A(\theta)|^2 - |A(\theta)|^4)A(\theta)}{1 - K\tau\exp(-i\Omega\tau)} + \frac{\sqrt{2D}\zeta(\theta)}{1 - K\tau\exp(-i\Omega\tau)}. \tag{4.10}$$

The corresponding stationary probability distribution for the amplitude reads

$$P(r) = Nr \exp\left(\frac{r^2}{D_{eff}}\left(\frac{\delta\lambda}{2} + \frac{sr^2}{4} - \frac{r^4}{6}\right)\right), \tag{4.11}$$

with

$$\delta\lambda = \lambda - K(1 + \cos(\Omega\tau)), \quad D_{eff} = \frac{D}{1 - K\tau\cos(\Omega\tau)}. \tag{4.12}$$

So we derived the probability distributions corresponding to the periodic solutions of the coupled system. To check the validity of our ansatz, we simulate the stochastic dynamics of the Eqs. (4.2, 4.3) and determine the stationary probability distribution for the amplitude. We will just study the in-phase solution, because for the anti-phase solution the results would not differ.

This comparison is made for three different coupling strengths: in Fig. 4.1 we set $K = 0.05$, and in Figs. 4.2 we have $K = 0.25$ and $K = 0.5$.

The analytical expression (Eq. (4.6)) displays the qualitative behaviour of the numerics for weak and strong noise intensities and quite good agreement in the low noise limit ($D \to 0$). But for the parameter values of interest, strictly speaking for intermediate noise strengths, (where a stochastic P-bifurcation was observed in the chapters 2 and 3), the analytical and numerical results differ so strongly that this ansatz does not seem to be useful, especially for higher coupling strengths. The best agreement is achieved for the weak coupling strength $K = 0.05$ in Fig. 4.1.

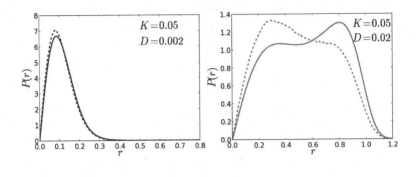

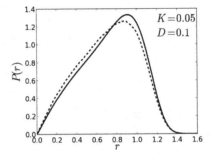

Figure 4.1: Stationary amplitude probability distribution of oscillator z_1, analytically (solid, Eq. (4.6)) and numerically (dashed, Eqs. (4.2, 4.3)) estimated. Parameters: $\tau = 1.0$, $\lambda = -0.26$, $\omega_0 = 2\pi$, $s = 1$.

For $D = 0$, the relation $z_S = z_1 = z_2$ is valid, but for $D \neq 0$ we have to write $z_S = z_1 \approx z_2$ because of the underlying stochastic dynamics. The main reason is probably that the deterministic ansatz ignores the stochastic input of the coupled oscillator. Because we chose two different white noise terms in the Eqs. (4.2, 4.3) a second independent noise source enters the equation. Equation (4.6) could be a proper description if the white noise terms are equal.

For small values of the coupling constant K, we can more or less neglect the stochastic influence of the coupled oscillator, therefore, we have the best agreement with the numerics. But by increasing K the coupled stochastic oscillator becomes more important for the dynamics.

Before we continue with our discussion, we use the numerical results to check whether a stochastic P-bifurcation takes place in the coupled oscillator system or not.

It turns out that a stochastic P-bifurcation takes place only for weak coupling (Fig. 4.3): in this case the curve for $D = 0.023$ represents a bimodal shaped distribution.

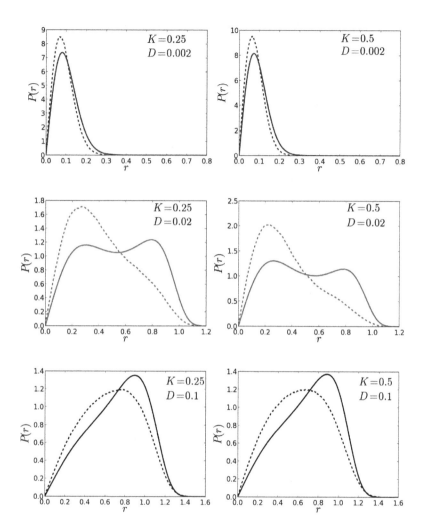

Figure 4.2: Stationary amplitude probability distribution of oscillator z_1, analytically (solid, Eq. (4.6)) and numerically (dashed, Eqs. (4.2, 4.3)) estimated. Parameters: $\tau = 1.0$, $\lambda = -0.26$, $\omega_0 = 2\pi$, $s = 1$.

For higher values of the coupling strength, the distribution is smeared out and stays unimodal (Fig. 4.4).
Probably the form of the coupling term could also have an influence on the results.

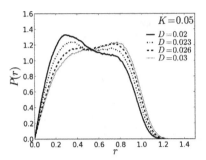

Figure 4.3: Stationary probability distribution for the amplitude of z_1, numerically calculated from the Eqs. (4.2, 4.3). Parameters: $\tau = 1.0$, $\lambda = -0.26$, $\omega_0 = 2\pi$ $s = 1$.

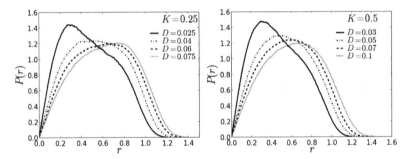

Figure 4.4: Stationary probability distribution for the amplitude of z_1, numerically calculated from Eqs. (4.2, 4.3). Parameters: $\tau = 1.0$, $\lambda = -0.26$, $\omega_0 = 2\pi$, $s = 1$.

We can rewrite Eq. (4.2)

$$\dot{z}_1(t) = (\underbrace{\lambda - K}_{=\eta} + i\omega_0 + s|z_1(t)|^2 - |z_1(t)|^4)z_1(t) + Kz_2(t - \tau) + \sqrt{2D}\xi_1(t) \quad (4.13)$$

where η represents an effective bifurcation parameter. By increasing the coupling strength, the oscillator z_1 is shifted away from the saddle-node bifurcation of limit cycles and, therefore, out of the parameter regime for a bimodal shaped probability distribution. Because the oscillator z_2 is driven by a different stochastic force, a different behaviour is observed than in the case with time-delayed feedback.

The single system exhibits a stochastic P-bifurcation for $\lambda \in (-0.33, -0.25)$. The values $K = 0.05$ and $\lambda = -0.26$ result in an effective value $\eta = -0.31$, where it is still possible to discover a stochastic P-bifurcation, see Fig. 4.3. For $K = 0.25$ and $\lambda = -0.26$ we obtain for $\eta = -0.51$ which is far away from the region of bimodality, see Fig. 4.4. In the case of higher coupling strength, the stochastic force of the other oscillator also plays an important role, which we will see in the investigation of coherence resonance.

Further studies are needed to understand the behaviour of the coupled stochastic oscillators.

4.2   Coherence resonance in coupled oscillators

Next, we simulate the stochastic dynamics of the coupled oscillators and calculate the correlation time for the real part of the oscillator z_1, which represents the stochastic in-phase solution for $\tau = 1$ and the stochastic anti-phase solution for $\tau = 0.5$. The numerical results are given for different values of the bifurcation parameters λ (Fig. 4.5) and for different coupling strengths (Fig. 4.6). Again only the in-phase solution is investigated because the results for the anti-phase solution do not differ (compare [74], Figs. 8b, 8e).

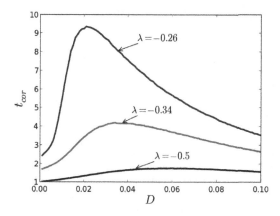

Figure 4.5: Numerical simulation of the correlation time t_{cor} for different
values of λ for the real part of z_1 (Eqs. (4.2, 4.3)).
Parameters: $\tau = 1.0$, $K = 0.5$, $\omega_0 = 2\pi$, $s = 1$.

Figure 4.5 displays the result, which we also obtained in the previous chapters: for the value of λ close to the saddle-node bifurcation of limit cycles, we get the most pronounced coherence resonance and by going further away the maximum of the correlation time t_{cor} decreases.

For different coupling strengths (Fig. 4.6) we obtain a result, which provides the suggestions made on the effect of the coupling term. For small values of K the correlation time decreases and then increases again by further increasing the coupling strength.

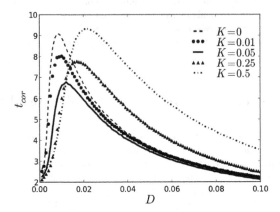

Figure 4.6: Numerical simulation of the correlation time t_{cor} for different
values of K for the real part of z_1 (Eqs. (4.2, 4.3)).
Parameters: $\tau = 1.0$, $\lambda = -0.26$, $\omega_0 = 2\pi$, $s = 1$.

The reason could be that for small coupling strengths the system is shifted away
form the parameter regime, where a pronounced coherence resonance can be observed. At a certain value of K the stochastic input from the coupled oscillator take
over the main influence on the dynamics and the two independent noise sources
play a constructive role, which results in an increase of the correlation time. The
maximum is shifted to higher values, which may be connected to the effective bifurcation parameter η (Eq. (4.13)), which increases to more negative values; as we
observed in Fig. 4.5 the maximum of the correlation time t_{cor} is shifted to higher
values of the noise intensity for more negative values of λ.
Note that for a coupling term of the form $K z_2(t - \tau)$ instead of $K(z_1 - z_2(t - \tau))$ it
was observed in [74] that an increase of the coupling strength leads to more regular
motion (smaller width at half maximum) and a smaller optimal noise intensity.
Although we observe a clear pronounced maximum for the correlation time at the
value $\lambda = -0.26$, there is no stochastic P-bifurcation present; the underlying distribution has a unimodal shape (Fig. 4.4).
Therefore, we calculate the ghost weight (Eq. (2.52))

$$g(D) = \int_{r_0}^{\infty} P(r)dr \qquad (4.14)$$

and the corresponding derivative due to the noise intensity to check if the stochastic phase space shows an influence of the ghost.
Although we used analytical expressions of the stationary probability distribution
for the calculation of the ghost weight in the previous chapters, we only have numerical data available here.
We calculate 100 realisations of the probability distribution for values of the noise

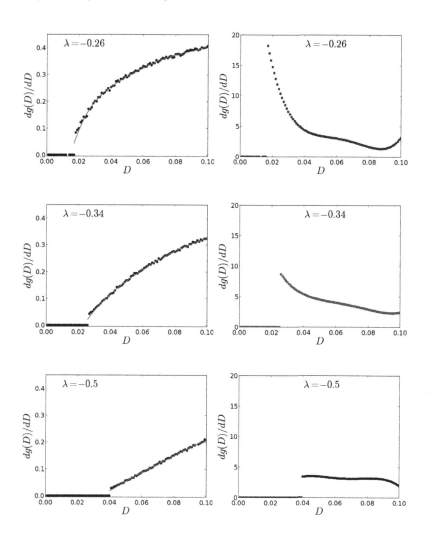

Figure 4.7: Ghost weight $g(D)$ (left) and the corresponding derivative $\frac{dg(D)}{dD}$ (right) for z_1 obtained from numerical data, represented by dots. The line shows the data fit for the values of $g(D) > 0$. In the right plot the dots show the corresponding derivative, calculated from the fit function.

Parameters: $K = 0.5$, $\tau = 1.0$, $\omega_0 = 2\pi$, $s = 1$.

intensity $D \in [0.001, 0.1]$. Because of the stochastic character of our equations, the

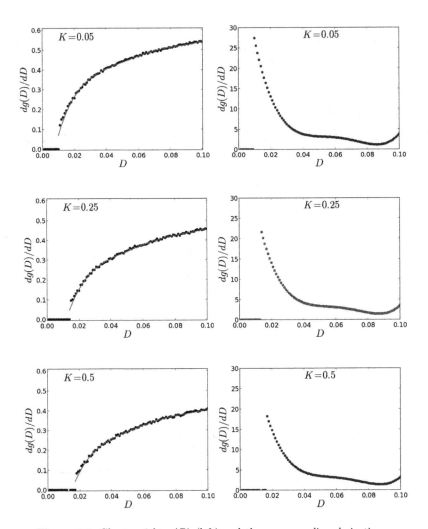

Figure 4.8: Ghost weight $g(D)$ (left) and the corresponding derivative $\frac{dg(D)}{dD}$ (right) for z_1 obtained from numerical data, represented by dots. The line shows the data fit for the values of $g(D) > 0$. In the right plot the dots show the corresponding derivative, calculated from the fit function.
Parameters: $\tau = 1.0$, $\omega_0 = 2\pi$, $\lambda = -0.26$, $s = 1$.

probability distributions are not smooth lines. Thus, we do not obtain a smooth transition for the ghost weight from $g(D) = 0$ to $g(D) \neq 0$. We use a polynomial function of order five as a fit function for the values $g(D) \neq 0$, including the last value $g(D) = 0$.

Then we calculate the derivative of the polynomial fit function and plot it as single points to illustrate that all data used for this calculation are available as discrete values. The values $g(D) = 0$ do not contribute for the derivative, so we set them zero, $\frac{d}{dD}g(D) = 0$. Hence, we have to interpret the first value $\frac{d}{dD}g(D) \neq 0$ as the maximum, which is in nearly all cases also the highest value.

As we observed in the sections 2.4.2 and 3.2.3, the derivative of the ghost weight decreases from the maximum value monotonically for $D \rightarrow 0.1$. Here, we find that for values $D \rightarrow 0.1$ the derivative shows unexpected behaviour, which is caused by the fit function. The fit function is just a suitable visualisation of the ghost weight up to values of the noise intensity $D = 0.1$.

We observe that for different values of λ, the most pronounced resonance-like behaviour of the derivative of the ghost weight $\frac{dg(D)}{dD}$ is shown close to the saddle-node bifurcation of limit cycles and decreases by going further away to more negative values of λ, see Fig. 4.7. This result was also observed for the single Hopf normal form, with and without time-delayed feedback, see sections 2.4.2, 3.2.3.

For different coupling strengths, we obtain the same result as for the Hopf normal form with self-feedback: having a small coupling strength, the maximum of the ghost weight derivative shows a greater value as for stronger coupling strengths, see Fig. 4.8. This is somehow connected again with the scaling of the noise strength (see Eqs. (3.37, 4.7, 4.12)) in contrast to the correlation time, where a higher coupling strength leads to a higher correlation time (in Fig. 4.6 for the values $K = 0.05$, $K = 0.25$ and $K = 0.5$).

Looking at the Figs. 4.3, 4.4, 4.8, it can be observed that the derivative of the ghost weight is most pronounced in the regime of bimodality and less pronounced in the region, where the probability distribution has a unimodal shape.

The ghost weight and the corresponding derivative show that the ghost of the stable limit cycle has a strong influence of the stochastic dynamics and that it is also possible to observe coherence resonance in the absence of a stochastic P-bifurcation.

Chapter 5

Conclusion

In this work, we have studied noise effects in non-excitable nonlinear systems with time-delayed feedback, and in delay-coupled oscillator systems. The nonlinear systems were represented by Hopf normal forms with additive Gaussian white noise. For a single system, we performed the bifurcation and stability analysis for the deterministic equation. Then we turned on the noise and derived the stationary probability distribution for the amplitude. By changing the noise intensity, a stochastic P-bifurcation occurred for the subcritical Hopf normal form. We showed that no stochastic P-bifurcation takes place in the supercritical case. Furthermore, we investigated coherence resonance in our system. Besides numerical simulations, we made use of statistical linearisation techniques to obtain analytical expressions for the measures of coherence (peak height, width at half maximum, signal-to-noise ratio, and correlation time). Our analytical approximations showed good agreement with the numerical simulations and we observed an extremum of the measures of coherence at an intermediate noise strength. The mechanism of coherence resonance in non-excitable systems can be related to stochastic P-bifurcations: a most regular motion is provided in the regime of a bimodal distribution. However, coherence resonance can also be observed in a much less pronounced form outside of this region. So we introduced the ghost weight as a novel measure for the explanation. The ghost weight measures the part of the distribution, which exceeds the radius, where the saddle-node bifurcation of limit cycles takes place in the deterministic case. The derivative of the ghost weight with respect to the noise intensity exhibits a maximum at an intermediate noise strength. In the regime of bimodality, this maximum is the highest, but also outside the region of bimodality it is clearly pronounced. So the ghost of the limit cycle has still a strong effect on the stochastic dynamics outside of the regime of a bimodal probability distribution.

The ghost weight was not introduced as a unique measure for the regularity of the noise-induced oscillations; it just shows the influence of the ghost of the stable limit cycle for the stochastic dynamics and provides an explanation, why coherence resonance can also be observed outside the region of bimodality.

Adding a time-delayed self-feedback term, we performed the steps of investigation similar to the non-delayed case. We started with a bifurcation and stability analysis for the deterministic delayed system. Switching the noise on, we were able to reduce the stochastic delay differential equation to an effective stochastic differential equation close to the Hopf bifurcation, using a multiple scaling perturbation approach. Hence, we calculated the stationary probability distribution for the amplitude and our approximation shows excellent agreement with the numerics in a wide parameter range. For half integer delays, we observed a suppression of the bimodality and for the integer values of the delay we obtained a change of the stochastic bifurcation diagram - strictly speaking a change of the parameter regime - where the bimodality can be found. We made again use of the statistical linearisation and derived an expression for the correlation time. Here, we also found that for half integer values of the time delay the correlation time decreased, which corresponds to a suppression of coherence resonance, whereas the correlation time increased for integer time delays, so coherence resonance is enhanced. Furthermore, the position of the optimal noise intensity is shifted to higher values of the noise strength for higher integer values of the time delay.

The enhancement and the suppression is related to the stability of the deterministic focus. The shift of the optimal noise intensity can be explained by the change of the regime of bimodality for the different time delays: a higher noise strength is needed to reach the parameter region, where the system has a bimodal shaped probability distribution. As already shown in chapter 2, the derivative of the ghost weight displayed again a maximum, also for parameter values outside the regime of bimodality. The derivative of the ghost weight was also calculated for different time delays and coupling strengths; these comparisons show that the ghost weight does not tell anything about the regularity of the noisy oscillations. The maximum decreased for higher coupling strengths and for higher integer values of the time delay, which could be related to the scaling of the noise intensity by the time-delayed coupling.

Two delay-coupled stochastic systems were investigated in chapter 4. Numerical simulations showed that a stochastic P-bifurcation took only place for small coupling strengths. But coherence resonance was observed, even in the absence of the stochastic P-bifurcation. Therefore, the derivative of the ghost weight was estimated, which showed a high maximum due to the noise intensity. An analytical approximation of the in-phase and anti-phase solutions of the two coupled systems failed to give a suitable expression of the probability distribution. The derivative of the ghost weight was investigated for different parameter values for the deterministic bifurcation parameter λ, but also for different coupling strengths K and different delay times τ.

Chapter 6

Outlook

In this thesis, we only studied a single stochastic oscillator and two coupled oscillators. It will be interesting now to see if we can extent our methods to other network motifs (three or four coupled oscillators with directed or undirected coupling) or even large networks to find similar results for some general statements. Different network topologies or coupling schemes could be studied to find out whether and how the noise effects are affected by the structure of the network. The studies on the ghost weight could be applied to larger systems to check, if it is still a suitable way to provide an explanation of the mechanism of coherence resonance. It is also necessary to prove the statement that coherence resonance is most pronounced in the regime of a bimodal probability distribution for coupled systems.

Another aspect for further investigations could be the development of a suitable approach of the probability distribution for higher time delays for the single oscillator with self-feedback and for coupled oscillator systems. For large networks, the methods presented in [100–103] might be useful to derive an analytical expression for the probability distribution. Otherwise the investigations on stochastic networks would be restricted to numerical simulations. The stochastic impact of the coupled oscillators could be modelled with delayed noise terms [104–107], or a description with coloured noise for the interplay of different stochastic forces might be helpful. We only made use of Gaussian white noise in our studies. The connection between other types of noise, e.g. coloured noise, and time delay could be investigated.

Stochastic D-bifurcations were not considered in our work. In the context of larger networks, it might be important to check how the stability is affected in the presence of many stochastic forces. The influence of time delay to this type of bifurcation could also play an interesting role [108].

Bibliography

[1] V. S. Anishchenko, V. Astakhov, A. B. Neiman, T. Vadivasova, and L. Schimansky-Geier: *Nonlinear dynamics of chaotic and stochastic systems: tutorial and modern developments* (Springer, Berlin, 2007).

[2] W. Horsthemke and R. Lefever: *Noise-Induced Transitions. Theory and Applications in Physics, Chemistry, and Biology* (Springer Verlag, Berlin, 1984).

[3] L. Arnold: *Random Dynamical Systems*, Springer Monographs in Mathematics (Springer, Berlin, 2003).

[4] L. Gammaitoni, P. Hänggi, P. Jung, and F. Marchesoni: *Stochastic resonance*, Rev. Mod. Phys. **70**, 223 (1998).

[5] G. Hu, T. Ditzinger, C. Z. Ning, and H. Haken: *Stochastic resonance without external periodic force*, Phys. Rev. Lett. **71**, 807 (1993).

[6] A. S. Pikovsky and J. Kurths: *Coherence resonance in a noise-driven excitable system*, Phys. Rev. Lett. **78**, 775 (1997).

[7] O. V. Ushakov, H. J. Wünsche, F. Henneberger, I. A. Khovanov, L. Schimansky-Geier, and M. A. Zaks: *Coherence resonance near a Hopf bifurcation*, Phys. Rev. Lett. **95**, 123903 (2005).

[8] T. Erneux: *Applied delay differential equations* (Springer, 2009).

[9] J. K. Hale and S. M. Verduyn Lunel: *Introduction to Functional Differential Equations* (Springer, New York, 1993).

[10] K. Pyragas: *Continuous control of chaos by self-controlling feedback*, Phys. Lett. A **170**, 421 (1992).

[11] E. Ott, C. Grebogi, and J. A. Yorke: *Controlling chaos*, Phys. Rev. Lett. **64**, 1196 (1990).

[12] E. Schöll and H. G. Schuster (Editors): *Handbook of Chaos Control* (Wiley-VCH, Weinheim, 2008), second completely revised and enlarged edition.

[13] P. Hövel and E. Schöll: *Control of unstable steady states by time-delayed feedback methods*, Phys. Rev. E **72**, 046203 (2005).

[14] T. Dahms, P. Hövel, and E. Schöll: *Control of unstable steady states by extended time-delayed feedback*, Phys. Rev. E **76**, 056201 (2007).

[15] W. Just, B. Fiedler, V. Flunkert, M. Georgi, P. Hövel, and E. Schöll: *Beyond odd number limitation: a bifurcation analysis of time-delayed feedback control*, Phys. Rev. E **76**, 026210 (2007).

[16] B. Fiedler, S. Yanchuk, V. Flunkert, P. Hövel, H. J. Wünsche, and E. Schöll: *Delay stabilization of rotating waves near fold bifurcation and application to all-optical control of a semiconductor laser*, Phys. Rev. E **77**, 066207 (2008).

[17] C. v. Loewenich, H. Benner, and W. Just: *Experimental relevance of global properties of time-delayed feedback control*, Phys. Rev. Lett. **93**, 174101 (2004).

[18] C. v. Loewenich, H. Benner, and W. Just: *Experimental verification of Pyragas-Schöll-Fiedler control*, Phys. Rev. E **82**, 036204 (2010).

[19] S. Schikora, H. J. Wünsche, and F. Henneberger: *Odd-number theorem: Optical feedback control at a subcritical Hopf bifurcation in a semiconductor laser*, Phys. Rev. E **83**, 026203 (2011).

[20] H. Nakajima: *On analytical properties of delayed feedback control of chaos*, Phys. Lett. A **232**, 207 (1997).

[21] B. Fiedler, V. Flunkert, M. Georgi, P. Hövel, and E. Schöll: *Refuting the odd number limitation of time-delayed feedback control*, Phys. Rev. Lett. **98**, 114101 (2007).

[22] A. Amann and E. W. Hooton: *An odd-number limitation of extended time-delayed feedback control in autonomous systems*, Phil. Trans. R. Soc. A **371**, 20120463 (2013).

[23] D. Strehober, E. Schöll, and S. H. L. Klapp: *Feedback control of flow alignment in sheared liquid crystals*, Phys. Rev. E **88**, 062509 (2013).

[24] W. Kopylov, C. Emary, E. Schöll, and T. Brandes: *Time-delayed feedback control of the Dicke-Hepp-Lieb superradiant quantum phase transition*, New J. Phys **17**, 013040 (2015).

[25] N. B. Janson, A. G. Balanov, and E. Schöll: *Delayed feedback as a means of control of noise-induced motion*, Phys. Rev. Lett. **93**, 010601 (2004).

[26] R. Aust, P. Hövel, J. Hizanidis, and E. Schöll: *Delay control of coherence resonance in type-I excitable dynamics*, Eur. Phys. J. ST **187**, 77 (2010).

[27] R. Albert and A. L. Barabasi: *Statistical mechanics of complex networks*, Rev. Mod. Phys. **74**, 47 (2002).

[28] M. E. J. Newman, A. L. Barabasi, and D. J. Watts: *The Structure and Dynamics of Networks* (Princeton University Press, 2006).

[29] C. U. Choe, T. Dahms, P. Hövel, and E. Schöll: *Controlling synchrony by delay coupling in networks: from in-phase to splay and cluster states*, Phys. Rev. E **81**, 025205(R) (2010).

[30] V. Flunkert, S. Yanchuk, T. Dahms, and E. Schöll: *Synchronizing distant nodes: a universal classification of networks*, Phys. Rev. Lett. **105**, 254101 (2010).

[31] J. Lehnert, T. Dahms, P. Hövel, and E. Schöll: *Loss of synchronization in complex neural networks with delay*, Europhys. Lett. **96**, 60013 (2011).

[32] A. Keane, T. Dahms, J. Lehnert, S. A. Suryanarayana, P. Hövel, and E. Schöll: *Synchronisation in networks of delay-coupled type-I excitable systems*, Eur. Phys. J. B **85**, 407 (2012).

[33] T. Dahms, J. Lehnert, and E. Schöll: *Cluster and group synchronization in delay-coupled networks*, Phys. Rev. E **86**, 016202 (2012).

[34] E. Schöll: *Synchronization in delay-coupled complex networks*, in *Advances in Analysis and Control of Time-Delayed Dynamical Systems* (World Scientific, Singapore, 2013), Ed. by J.-Q. Sun, Q. Ding, chap. 4, pp. 57–83.

[35] S. Heiligenthal, T. Dahms, S. Yanchuk, T. Jüngling, V. Flunkert, I. Kanter, E. Schöll, and W. Kinzel: *Strong and weak chaos in nonlinear networks with time-delayed couplings*, Phys. Rev. Lett. **107**, 234102 (2011).

[36] V. Flunkert and E. Schöll: *Chaos synchronization in networks of delay-coupled lasers: Role of the coupling phases*, New. J. Phys. **14**, 033039 (2012).

[37] B. Fiedler, V. Flunkert, P. Hövel, and E. Schöll: *Delay stabilization of periodic orbits in coupled oscillator systems*, Phil. Trans. R. Soc. A **368**, 319 (2010).

[38] C. U. Choe, H. Jang, V. Flunkert, T. Dahms, P. Hövel, and E. Schöll: *Stabilization of periodic orbits near a subcritical Hopf bifurcation in delay-coupled networks*, Dyn. Sys. **28**, 15 (2013).

[39] B. Hauschildt, N. B. Janson, A. G. Balanov, and E. Schöll: *Noise-induced cooperative dynamics and its control in coupled neuron models*, Phys. Rev. E **74**, 051906 (2006).

[40] P. Hövel, M. A. Dahlem, and E. Schöll: *Control of synchronization in coupled neural systems by time-delayed feedback*, Int. J. Bifur. Chaos **20**, 813 (2010).

[41] R. Vicente, L. L. Gollo, C. R. Mirasso, I. Fischer, and P. Gordon: *Dynamical relaying can yield zero time lag neuronal synchrony despite long conduction delays*, Proc. Natl. Acad. Sci. U.S.A. **105**, 17157 (2008).

[42] E. Schöll, G. Hiller, P. Hövel, and M. A. Dahlem: *Time-delayed feedback in neurosystems*, Phil. Trans. R. Soc. A **367**, 1079 (2009).

[43] V. Flunkert, O. D'Huys, J. Danckaert, I. Fischer, and E. Schöll: *Bubbling in delay-coupled lasers*, Phys. Rev. E **79**, 065201 (R) (2009).

[44] K. Hicke, O. D'Huys, V. Flunkert, E. Schöll, J. Danckaert, and I. Fischer: *Mismatch and synchronization: Influence of asymmetries in systems of two delay-coupled lasers*, Phys. Rev. E **83**, 056211 (2011).

[45] M. C. Soriano, J. García-Ojalvo, C. R. Mirasso, and I. Fischer: *Complex photonics: Dynamics and applications of delay-coupled semiconductors lasers*, Rev. Mod. Phys. **85**, 421 (2013).

[46] A. L. Hodgkin and A. F. Huxley: *A quantitative description of membrane current and its application to conduction and excitation in nerve*, J. Physiol. **117**, 500 (1952).

[47] R. FitzHugh: *Impulses and physiological states in theoretical models of nerve membrane*, Biophys. J. **1**, 445 (1961).

[48] J. Nagumo, S. Arimoto, and S. Yoshizawa.: *An active pulse transmission line simulating nerve axon.*, Proc. IRE **50**, 2061 (1962).

[49] B. Lindner, J. García-Ojalvo, A. B. Neiman, and L. Schimansky-Geier: *Effects of noise in excitable systems*, Phys. Rep. **392**, 321 (2004).

[50] B. van der Pol: *On relaxation oscillations*, Phil. Mag. **2**, 978 (1926).

[51] S. H. Strogatz: *Nonlinear Dynamics and Chaos* (Westview Press, Cambridge, MA, 1994).

[52] J. Pomplun, A. Amann, and E. Schöll: *Mean field approximation of time-delayed feedback control of noise-induced oscillations in the Van der Pol system*, Europhys. Lett. **71**, 366 (2005).

[53] C. W. Gardiner: *Handbook of Stochastic Methods for Physics, Chemistry and the Natural Sciences* (Springer, Berlin, 2002).

[54] N. G. van Kampen: *Stochastic Processes in Physics and Chemistry* (North-Holland, Amsterdam, 2003).

[55] H. Risken: *The Fokker-Planck Equation* (Springer, Berlin, 1996), 2nd ed.

[56] W. Ebeling, H. Herzel, W. Richert, and L. Schimansky-Geier: *Influence of noise on Duffing-van der Pol oscillators*, J. Appl. Math. Mech. (ZAMM) **66**, 141 (1986).

[57] L. Schimansky-Geier and H. Herzel: *Positive Lyapunov exponents in the Kramers oscillator*, J. Stat. Phys. **70**, 141 (1993).

[58] R. Erban, S. J. Chapman, I. G. Kevrekidis, and T. Vejchodsky: *Analysis of a stochastic chemical system close to a SNIPER bifurcation of its mean-field model*, SIAM J. Appl. Math. **70**, 984 (2009).

[59] A. Zakharova, J. Kurths, T. Vadivasova, and A. Koseska: *Analysing dynamical behavior of cellular networks via stochastic bifurcations*, PLoS ONE **6**, e19696 (2011).

[60] L. Billings, I. B. Schwartz, D. S. Morgan, E. M. Bollt, R. Meucci, and E. Allaria: *Stochastic bifurcation in driven laser systems: Experiment and theory*, Phys. Rev. E **70**, 026220 (2004).

[61] A. Zakharova, T. Vadivasova, V. Anishchenko, A. Koseska, and J. Kurths: *Stochastic bifurcations and coherencelike resonance in a self-sustained bistable noisy oscillator*, Phys. Rev. E **81**, 011106 (2010).

[62] A. Zakharova, A. Feoktistov, T. Vadivasova, and E. Schöll: *Coherence resonance and stochastic synchronization in a nonlinear circuit near a subcritical Hopf bifurcation*, Eur. Phys. J. Spec. Top. **222**, 2481 (2013).

[63] P. M. Geffert, A. Zakharova, A. Vüllings, W. Just, and E. Schöll: *Modulating coherence resonance in non-excitable systems by time-delayed feedback*, Eur. Phys. J. B **87**, 291 (2014).

[64] Y. Xu, R. Gu, H. Zhang, W. Xu, and J. Duan: *Stochastic bifurcations in a bistable Duffing-van der Pol oscillator with colored noise*, Phys. Rev. E **83**, 056215 (2011).

[65] K. Wiesenfeld: *Noisy precursors of nonlinear instabilities*, J. Stat. Phys. **38**, 1071 (1985).

[66] A. B. Neiman, P. I. Saparin, and L. Stone: *Coherence resonance at noisy precursors of bifurcations in nonlinear dynamical systems*, Phys. Rev. E **56**, 270 (1997).

[67] D. Ziemann, R. Aust, B. Lingnau, E. Schöll, and K. Lüdge: *Optical injection enables coherence resonance in quantum-dot lasers*, Europhys. Lett. **103**, 14002 (2013).

[68] C. Otto, B. Lingnau, E. Schöll, and K. Lüdge: *Manipulating coherence resonance in a quantum dot semiconductor laser via electrical pumping*, Opt. Express **22**, 13288 (2014).

[69] B. Lindner and L. Schimansky-Geier: *Analytical approach to the stochastic FitzHugh-Nagumo system and coherence resonance*, Phys. Rev. E **60**, 7270 (1999).

[70] R. L. Stratonovich: *Topics in the Theory of Random Noise*, vol. 1 (Gordon and Breach, New York, 1963).

[71] J. Kottalam, K. Lindenberg, and B. J. West: *Statistical replacement for systems with delta-correlated fluctuations*, J. Stat. Phys. **42**, 979 (1986).

[72] V. Flunkert, P. Hövel, and E. Schöll: *Coherence resonance - a mean field approach*, Poster at the Workshop Constructive Role of Noise in Complex Systems (2006).

[73] P. M. Geffert, A. Vüllings, V. Flunkert, and E. Schöll: *Mean-field approach to coherence resonance in Hopf oscillators* (2012), unpublished notes.

[74] A. Vüllings, E. Schöll, and B. Lindner: *Spectra of delay-coupled heterogeneous noisy nonlinear oscillators*, Eur. Phys. J. B **87**, 31 (2014).

[75] R. M. Corless, G. H. Gonnet, D. E. G. Hare, D. J. Jeffrey, and D. E. Knuth: *On the Lambert W function*, Adv. Comput. Math **5**, 329 (1996).

[76] S. A. Brandstetter, M. A. Dahlem, and E. Schöll: *Interplay of time-delayed feedback control and temporally correlated noise in excitable systems*, Phil. Trans. R. Soc. A **368**, 391 (2010).

[77] E. Schöll, A. G. Balanov, N. B. Janson, and A. B. Neiman: *Controlling stochastic oscillations close to a Hopf bifurcation by time-delayed feedback*, Stoch. Dyn. **5**, 281 (2005).

[78] C. Masoller: *Noise-induced resonance in delayed feedback systems*, Phys. Rev. Lett. **88**, 034102 (2002).

[79] V. Flunkert and E. Schöll: *Suppressing noise-induced intensity pulsations in semiconductor lasers by means of time-delayed feedback*, Phys. Rev. E **76**, 066202 (2007).

[80] G. Stegemann, A. G. Balanov, and E. Schöll: *Delayed feedback control of stochastic spatiotemporal dynamics in a resonant tunneling diode*, Phys. Rev. E **73**, 016203 (2006).

[81] J. Hizanidis and E. Schöll: *Control of coherence resonance in semiconductor superlattices*, Phys. Rev. E **78**, 066205 (2008).

[82] M. Kehrt, P. Hövel, V. Flunkert, M. A. Dahlem, P. Rodin, and E. Schöll: *Stabilization of complex spatio-temporal dynamics near a subcritical Hopf bifurcation by time-delayed feedback*, Eur. Phys. J. B **68**, 557 (2009).

[83] G. C. Sethia, J. Kurths, and A. Sen: *Coherence resonance in an excitable system with time delay*, Phys. Lett. A **364**, 227 (2007).

[84] J. Pomplun: *Time-delayed feedback control of noise-induced oscillations*, Master's thesis, TU Berlin (2005).

[85] U. Küchler and B. Mensch: *Langevin stochastic differential equation extended by a time delayed term*, Stoch. Rep **40**, 23 (1992).

[86] S. Guillouzic, I. L'Heureux, and A. Longtin: *Small delay approximation of stochastic delay differential equations*, Phys. Rev. E **59**, 3970 (1999).

[87] T. D. Frank and P. J. Beek: *Stationary solutions of linear stochastic delay differential equations: Applications to biological systems*, Phys. Rev. E **64**, 021917 (2001).

[88] T. D. Frank: *Delay Fokker-Planck equations, Novikov's theorem, and Boltz-mann distributions as small delay approximations*, Phys. Rev. E **72**, 011112 (2005).

[89] T. D. Frank, P. J. Beck, and R. Friedrich: *Fokker-Planck perspective on stochastic delay systems: Exact solutions and data analysis of biological systems*, Phys. Rev. E **68**, 021912 (2003).

[90] A. Amann, E. Schöll, and W. Just: *Some basic remarks on eigenmode expansions of time-delay dynamics*, Physica A **373**, 191 (2007).

[91] W. Just, F. Matthäus, and H. Sauermann: *On the degenerated soft-mode instability*, J. Phys. A: Math. Gen. **31** (1998).

[92] W. Just, H. Benner, and C. v. Loewenich: *On global properties of time-delayed feedback control: weakly nonlinear analysis*, Physica D **199**, 33 (2004).

[93] M. Gaudreault, A. M. Levine, and J. Viñals: *Pitchfork and Hopf bifurcation thresholds in stochastic equations with delayed feedback*, Phys. Rev. E **80**, 061920 (2009).

[94] M. Gaudreault, F. Drolet, and J. Viñals: *Analytical determination of the bifurcation thresholds in stochastic differential equations with delayed feedback*, Phys. Rev. E **82**, 051124 (2010).

[95] M. Gaudreault, F. Drolet, and J. Viñals: *Bifurcation threshold of the delayed van der Pol oscillator under stochastic modulation*, Phys. Rev. E **85**, 056214 (2012).

[96] I. Schneider: *Delayed feedback control of three diffusively coupled Stuart-Landau oscillators: a case study in equivariant Hopf bifurcation*, Phil. Trans. R. Soc. A **371**, 20120472 (2013).

[97] A. Vüllings: *Stochastic dynamics in networks with delay*, Master's thesis, HU Berlin (2012).

[98] S. Yanchuk and P. Perlikowski: *Delay and periodicity*, Phys. Rev. E **79**, 046221 (2009).

[99] O. D'Huys, R. Vicente, T. Erneux, J. Danckaert, and I. Fischer: *Synchronization properties of network motifs: Influence of coupling delay and symmetry*, Chaos **18**, 037116 (2008).

[100] D. Huber and L. S. Tsimring: *Dynamics of an ensemble of noisy bistable elements with global time delayed coupling*, Phys. Rev. Lett. **91**, 260601 (2003).

[101] D. Huber and L. S. Tsimring: *Cooperative dynamics in a network of stochastic elements with delayed feedback*, Phys. Rev. E **71**, 036150 (2005).

[102] T. D. Frank: *Nonlinear Fokker-Planck Equations: Fundamentals and Applications* (Springer, 2005).

[103] A. Pototsky and N. B. Janson: *Synchronization of a large number of continuous one-dimensional stochastic elements with time delayed mean field coupling*, Physica D **238**, 175 (2009).

[104] T. D. Frank: *Delay Fokker-Planck equations, perturbation theory, and data analysis for nonlinear stochastic systems with time delays*, Phys. Rev. E **71**, 031106 (2005).

[105] M. Borromeo, S. Giusepponi, and F. Marchesoni: *Recycled noise rectification: An automated Maxwell's daemon*, Phys. Rev. E **74**, 031121 (2006).

[106] M. Borromeo and F. Marchesoni: *Stochastic synchronization via noise recycling*, Phys. Rev. E **75**, 041106 (2007).

[107] A. Chéagé Chamgoué, R. Yamapi, and P. Woafo: *Dynamics of a biological system with time-delayed noise*, Eur. Phys. J. Plus **127**, 59 (2012).

[108] N. J. Ford and S. J. Norton: *Numerical investigation of D-bifurcations for a stochastic delay logistic equation*, Stoch. Dyn. **05**, 211 (2005).

Appendix A

Derivation of the Fokker-Planck equation

In general, a n-dimensional stochastic differential equation is written

$$d\mathbf{x} = \mathbf{A}(x,t)dt + \mathbf{B}(x,t)d\mathbf{W}(t), \tag{A.1}$$

where $\mathbf{x}$ is a state vector, $\mathbf{A}$ denotes the drift vector, $\mathbf{B}$ is the diffusion matrix, and $\mathbf{W}$(t) describes a Wiener process. The corresponding Fokker-Planck equation reads (we use the Stratonovich interpretation of a stochastic differential equation [53], which is usually used in physics)

$$\partial_t P(x,t) = -\sum_i \partial_i (A_i P(x,t)) + \frac{1}{2}\sum_{i,j,k} \partial_i (B_{ik}\partial_j [B_{jk} P(x,t)]), \tag{A.2}$$

where $P(x,t)$ is the probability density. In chapter 2, our equation for the Hopf normal form with Gaussian white noise $\xi(t) \in \mathbb{C}$ reads

$$\dot{z}(t) = (\lambda + i\omega_0 - a|z(t)|^2 - b|z(t)|^4)z(t) + \sqrt{2D}\xi(t). \tag{A.3}$$

$z(t)$ is the complex variable, λ denotes the bifurcation parameter, and ω_0 is the intrinsic frequency of the system. The real parameters a and b are used to distinguish between the supercritical ($a = 1$, $b = 0$) and the subcritical ($a = -1$, $b = 1$) Hopf normal form and $D \geq 0$ describes the strength of the fluctuations (noise strength). We first transform $z(t)$ into polar coordinates $z = re^{i\phi}$ and decompose the resulting equation into real and imaginary part. Also the noise variable is decomposed up into radius and phase part. The equation reads

$$\begin{pmatrix} \dot{r} \\ \dot{\phi} \end{pmatrix} = \underbrace{\begin{pmatrix} \lambda r - ar^3 - br^5 \\ \omega_0 \end{pmatrix}}_{=:\ \mathbf{A}} + \underbrace{\sqrt{2D} \begin{pmatrix} \cos(\phi) & \sin(\phi) \\ -\frac{\sin(\phi)}{r} & \frac{\cos(\phi)}{r} \end{pmatrix}}_{=:\ \mathbf{B}} \begin{pmatrix} \xi_r \\ \xi_\phi \end{pmatrix}. \tag{A.4}$$

For simplicity, we write $P(x,t) = P(x_1, x_2, t) = P(r, \phi, t) = P$ in the whole calculation. The first part of equation (A.2) for our system (eq.(A.4)) is

$$\sum_i \partial_i (A_i P(x,t)) = \partial_1 (A_1 P) + \partial_2 (A_2 P)$$

$$= \partial_r \left((\lambda r - ar^3 - br^5) P \right) + \partial_\phi (\omega_0 P). \quad (A.5)$$

The second part of equation (A.2) reads (we ignore the prefactors $\frac{1}{2}$ (eq.(A.2)) and $\sqrt{2D}$ (eq.(A.4)) for a moment)

$$\sum_{i,j,k} \partial_i (B_{ik} \partial_j [B_{jk} P(x,t)]) = \partial_1 (B_{11} \partial_1 [B_{11} P]) + \partial_2 (B_{21} \partial_1 [B_{11} P])$$

$$+ \partial_1 (B_{11} \partial_2 [B_{21} P]) + \partial_2 (B_{21} \partial_2 [B_{21} P])$$
$$+ \partial_1 (B_{12} \partial_1 [B_{12} P]) + \partial_2 (B_{22} \partial_1 [B_{12} P])$$
$$+ \partial_1 (B_{12} \partial_2 [B_{22} P]) + \partial_2 (B_{22} \partial_2 [B_{22} P]), \quad (A.6)$$

$$\partial_1 (B_{11} \partial_1 [B_{11} P]) = \partial_r (\cos(\phi) \partial_r [\cos(\phi) P])$$
$$= \cos^2(\phi) \partial_{rr} P, \quad (A.7)$$

$$\partial_2 (B_{21} \partial_1 [B_{11} P]) = \partial_\phi \left(\frac{-\sin(\phi)}{r} \partial_r (\cos(\phi) P) \right)$$
$$= -\frac{\cos^2(\phi)}{r} \partial_r P + \frac{\sin^2(\phi)}{r} \partial_r P - \frac{\sin(\phi) \cos(\phi)}{r} \partial_\phi \partial_r P, \quad (A.8)$$

$$\partial_1 (B_{11} \partial_2 [B_{21} P]) = \partial_r \left(\cos(\phi) \partial_\phi \left(\frac{-\sin(\phi)}{r} P \right) \right)$$
$$= \cos(\phi) \partial_r \left(\frac{-\cos(\phi)}{r} P - \frac{\sin(\phi)}{r} \partial_\phi P \right)$$
$$= \frac{\cos^2(\phi)}{r^2} P - \frac{\cos^2(\phi)}{r} \partial_r P$$
$$+ \frac{\cos(\phi) \sin(\phi)}{r^2} \partial_\phi P - \frac{\cos(\phi) \sin(\phi)}{r} \partial_r \partial_\phi P, \quad (A.9)$$

$$\partial_2 (B_{21} \partial_2 [B_{21} P]) = \partial_\phi \left(\frac{-\sin(\phi)}{r} \partial_\phi \left(\frac{-\sin(\phi)}{r} P \right) \right)$$
$$= \partial_\phi \left(\frac{-\sin(\phi)}{r} \left(\frac{-\cos(\phi)}{r} P - \frac{\sin(\phi)}{r} \partial_\phi P \right) \right)$$
$$= \frac{\cos^2(\phi)}{r^2} P - \frac{\sin^2(\phi)}{r^2} P$$
$$+ 3 \frac{\sin(\phi) \cos(\phi)}{r^2} \partial_\phi P + \frac{\sin^2(\phi)}{r^2} \partial_{\phi\phi} P, \quad (A.10)$$

$$\partial_1(B_{12}\partial_1[B_{12}P]) = \partial_r\left(\sin(\phi)\partial_r\left(\sin(\phi)P\right)\right)$$
$$= \sin^2(\phi)\partial_{rr}P, \qquad (A.11)$$

$$\partial_2(B_{22}\partial_1[B_{12}P]) = \partial_\phi\left(\frac{\cos(\phi)}{r}\partial_r\left(\sin(\phi)P\right)\right)$$
$$= -\frac{\sin^2(\phi)}{r}\partial_r P + \frac{\cos^2(\phi)}{r}\partial_r P + \frac{\sin(\phi)\cos(\phi)}{r}\partial_\phi\partial_r P, \quad (A.12)$$

$$\partial_1(B_{12}\partial_2[B_{22}P]) = \partial_r\left(\sin(\phi)\partial_r\left(\frac{\cos(\phi)}{r}P\right)\right)$$
$$= \sin(\phi)\partial_r\left(-\frac{\sin(\phi)}{r}P + \frac{\cos(\phi)}{r}\partial_\phi P\right)$$
$$= \frac{\sin^2(\phi)}{r^2}P + \frac{\sin(\phi)\cos(\phi)}{r}\partial_r\partial_\phi P$$
$$- \frac{\sin^2(\phi)}{r^2}\partial_r P - \frac{\sin(\phi)\cos(\phi)}{r^2}\partial_\phi P, \qquad (A.13)$$

$$\partial_2(B_{22}\partial_2[B_{22}P]) = \partial_\phi\left(\frac{\cos(\phi)}{r}\partial_\phi\left(\frac{\cos(\phi)}{r}P\right)\right)$$
$$= \partial_\phi\left(\frac{\cos(\phi)}{r}\left(-\frac{\sin(\phi)}{r}P + \frac{\cos(\phi)}{r}\partial_\phi P\right)\right)$$
$$= -\frac{\cos^2(\phi)}{r^2}P + \frac{\sin^2(\phi)}{r^2}P$$
$$- 3\frac{\sin(\phi)\cos(\phi)}{r^2}\partial_\phi P + \frac{\cos^2(\phi)}{r^2}\partial_{\phi\phi}P. \qquad (A.14)$$

Combining the equation (A.7) and equation (A.11) yields

$$\cos^2(\phi)\partial_{rr}P + \sin^2(\phi)\partial_{rr}P = \partial_{rr}P. \qquad (A.15)$$

Equation (A.10) and equation (A.14) result in

$$\frac{\cos^2(\phi)}{r^2}P - \frac{\sin^2(\phi)}{r^2}P + 3\frac{\sin(\phi)\cos(\phi)}{r^2}\partial_\phi P + \frac{\sin^2(\phi)}{r^2}\partial_{\phi\phi}P$$
$$- \frac{\cos^2(\phi)}{r^2}P + \frac{\sin^2(\phi)}{r^2}P - 3\frac{\sin(\phi)\cos(\phi)}{r^2}\partial_\phi P + \frac{\cos^2(\phi)}{r^2}\partial_{\phi\phi}P$$
$$= \frac{1}{r^2}\partial_{\phi\phi}P. \qquad (A.16)$$

The equations (A.8) and (A.12) vanish, when added together.
Equation (A.9) and equation (A.13) give

$$\frac{\cos^2(\phi)}{r^2}P - \frac{\cos^2(\phi)}{r}\partial_r P + \frac{\cos(\phi)\sin(\phi)}{r^2}\partial_\phi P - \frac{\cos(\phi)\sin(\phi)}{r}\partial_r\partial_\phi P$$
$$+ \frac{\sin^2(\phi)}{r^2}P + \frac{\sin(\phi)\cos(\phi)}{r}\partial_r\partial_\phi P - \frac{\sin^2(\phi)}{r^2}\partial_r P - \frac{\sin(\phi)\cos(\phi)}{r^2}\partial_\phi P$$
$$= \frac{1}{r^2}P - \frac{1}{r^2}\partial_r P. \tag{A.17}$$

Now we put the prefactors $\frac{1}{2}$ and $\sqrt{2D}$ back to our equations; we finally obtain from the equations (A.15), (A.16), and (A.17)

$$\rightarrow \quad D\left(\partial_{rr}P + \frac{1}{r^2}\partial_{\phi\phi}P + \frac{1}{r^2}P - \frac{1}{r^2}\partial_r P\right). \tag{A.18}$$

Combining equation (A.5) and equation (A.18), we end up with the Fokker-Planck equation for the noisy Hopf normal form (A.3)

$$\partial_t P = -\partial_r\left((\lambda r - ar^3 - br^5)P\right) - \partial_\phi\left(\omega_0 P\right)$$
$$+ D\left(\partial_{rr}P + \frac{1}{r^2}\partial_{\phi\phi}P + \frac{1}{r^2}P - \frac{1}{r^2}\partial_r P\right)$$
$$= \partial_r\left(\left(-\lambda r + ar^3 + br^5 - \frac{D}{r}\right)P + D\partial_r P\right) + \partial_\phi\left(-\omega_0 P + \frac{D}{r^2}\partial_\phi P\right). \tag{A.19}$$

Appendix B

Stochastic bifurcation diagram

The probability distribution (see eq.(2.21))

$$P(r) = Nr \exp\left(\frac{r^2}{D}\left(\frac{\lambda}{2} + \frac{r^2}{4} - \frac{r^4}{6}\right)\right) \tag{B.1}$$

undergoes a stochastic P-bifurcation by varying the noise intensity D. Here, we want to show the detailed calculation for the bifurcation lines.
The exponent of $P(r)$

$$\ln P = \frac{r^2}{D}\left(\frac{\lambda}{2} + \frac{r^2}{4} - \frac{r^4}{6}\right) + \ln(r) \tag{B.2}$$

shows an inflection point, which means

$$\frac{d}{dr}\ln(P(r)) = 0, \quad \frac{d^2}{dr^2}\ln(P(r)) = 0. \tag{B.3}$$

Therefore, we obtain two conditions

$$
\begin{aligned}
-D/r - \lambda r - r^3 + r^5 &= 0 \\
D/r^2 - \lambda - 3r^2 + 5r^4 &= 0.
\end{aligned}
\tag{B.4}
$$

By rewriting, we find two equations of third order for the variable r^2:

$$
\begin{aligned}
-D - \lambda r^2 - r^4 + r^6 &= 0 \mid I \\
D - \lambda r^2 - 3r^4 + 5r^6 &= 0 \mid II.
\end{aligned}
\tag{B.5}
$$

Now we reduce the high order to obtain a condition for the variable r^2

$$
\begin{array}{rcll}
r^6 - r^4 - \lambda r^2 - D & = & 0 & | \ I + II = Ia \\
5r^6 - 3r^4 - \lambda r^2 + D & = & 0 & | \ II,
\end{array}
\tag{B.6}
$$

$$
\begin{array}{rcll}
6r^6 - 4r^4 - 2\lambda r^2 & = & 0 & | \ Ia \cdot \dfrac{5}{2} \\
5r^6 - 3r^4 - \lambda r^2 + D & = & 0 & | \ II \cdot 3,
\end{array}
\tag{B.7}
$$

$$
\begin{array}{rcll}
15r^6 - 10r^4 - 5\lambda r^2 & = & 0 & | \ Ia \\
15r^6 - 9r^4 - 3\lambda r^2 + 3D & = & 0 & | \ II - Ia = IIa,
\end{array}
\tag{B.8}
$$

$$
\begin{array}{rcll}
3r^4 - 2r^2 - \lambda & = & 0 & | \ (Ia) \cdot b = Ib \\
r^4 + 2\lambda r^2 + 3D & = & 0 & | \ (IIa) \cdot 3c = IIb,
\end{array}
\tag{B.9}
$$

$$
\begin{array}{rcll}
3r^4 - 2r^2 - \lambda & = & 0 & | \ Ib \\
3r^4 + 6\lambda r^2 + 9D & = & 0 & | \ IIb - Ib,
\end{array}
\tag{B.10}
$$

$$
(6\lambda + 2)r^2 + 9D + \lambda = 0 \quad | \ IIb,
\tag{B.11}
$$

$$
r^2 = -\frac{9D + \lambda}{6\lambda + 2} > 0.
\tag{B.12}
$$

This shows that only values of $\lambda < 0$ are allowed; otherwise the radius becomes imaginary. Inserting this condition to the equations (B.5), we find after some calculation (or using mathematica)

$$
\frac{(2 - 27D + 9\lambda)(-D(4 + 27D) - 18D\lambda + \lambda^2 + 4\lambda^3)}{8(1 + 3\lambda)^3} = 0,
\tag{B.13}
$$

$$
\frac{(-2 + 135D + 9\lambda)(-D(4 + 27D) - 18D\lambda + \lambda^2 + 4\lambda^3)}{8(1 + 3\lambda)^3} = 0.
\tag{B.14}
$$

The common condition from this two equations is

$$
\begin{aligned}
0 &= -D(4 + 27D) - 18D\lambda + \lambda^2 + 4\lambda^3 \\
&= -27D^2 - D(4 - 18\lambda) + \lambda^2 + 4\lambda^3.
\end{aligned}
\tag{B.15}
$$

This result also follows directly by calculating the resultant of the two polynomials. Solving equation (B.15) to D, we obtain

$$
D_{1,2} = \frac{1}{27}\left(-9\lambda - 2\left(1 \pm \sqrt{(1 + 3\lambda)^3}\right)\right).
\tag{B.16}
$$

Equation (B.16) represents the two bifurcation lines (see also section 2.3, figure 2.3).

Appendix C

Eigenmode expansion

Here we want to collect the main steps from [90] to understand the calculation progress in chapter 3 (see eq.(3.41)).
We start with a linear delay differential equation

$$\dot{x}(t) = -ax(t) + bx(t - \tau), \qquad (C.1)$$

with the constant coefficients a and b and the initial condition

$$x(\theta) = \phi(\theta), \quad -\tau \le \theta \le 0. \qquad (C.2)$$

We are interested in solutions of exponential type, $x(t) = \exp(\lambda t)$. The corresponding characteristic equation reads

$$\lambda = -a + b\exp(-\lambda\tau). \qquad (C.3)$$

We can solve this equation by using the Lambert W-function

$$\lambda_l = -a + \frac{W_l(b\tau\exp(a\tau))}{\tau}, \qquad (C.4)$$

where W_l denotes the l-branch of the Lambert W-function. The solution of equation (C.1) can be written as

$$x(t) = \sum_l c_l \exp(\lambda_l t). \qquad (C.5)$$

The coefficients c_l are determined in agreement with the initial condition

$$\phi(\theta) = \sum_l c_l \underbrace{\exp(\lambda_l \theta)}_{=U_l(\theta)}. \qquad (C.6)$$

We use the condition (for ($l \neq l'$) and due to equation (C.3))

$$b\exp(-\lambda_{l'}\tau)\int_{-\tau}^{0}\exp(-\lambda_{l'}\theta)\exp(-\lambda_{l}\theta)d\theta = \frac{b\exp(-\lambda_{l'}\tau) - b\exp(-\lambda_{l}\tau)}{\lambda_{l} - \lambda_{l'}}$$

$$= -1 \qquad\qquad (C.7)$$

to show that the expression

$$V_{l}^{*}(\theta) = [\delta(\theta) + b\exp(-\lambda_{l}(\theta + \tau))]/N_{l} \qquad\qquad (C.8)$$

is orthogonal to $U_{l}(\theta)$ (eq.(C.6))

$$N_{l}\int_{-\tau}^{0}V_{l'}^{*}(\theta)U_{l}(\theta)d\theta = 1 + b\exp(-\lambda_{l'}\tau)\int_{-\tau}^{0}\exp(-\lambda_{l'}\theta)\exp(-\lambda_{l}\theta)d\theta$$

$$= 0 \qquad\qquad (C.9)$$

We choose the normalisation constant N_{l} that

$$1 = \int_{-\tau}^{0}V_{l}^{*}(\theta)U_{l}(\theta)d\theta \qquad\qquad (C.10)$$

is valid. So we have

$$N_{l} = 1 + b\tau\exp(-\lambda\tau) = 1 + a\tau + \lambda_{l}\tau \qquad\qquad (C.11)$$

and hence, we can estimate the coefficients c_{l}

$$\int_{-\tau}^{0}V_{l}^{*}(\theta)\phi(\theta)d\theta = c_{l}. \qquad\qquad (C.12)$$

Finally the solution of equation (C.1) reads

$$x(t) = \sum_{l}\left[\phi(0) + \int_{-\tau}^{0}b\exp(-\lambda_{l}(\theta + \tau))\phi(\theta)d\theta\right]\frac{\exp(\lambda_{l}t)}{N_{l}}$$

$$= T(t,0)\phi(0) + \int_{-\tau}^{0}T(t,\theta + \tau)b\phi(\theta)d\theta, \qquad\qquad (C.13)$$

with

$$T(t,t') = \begin{cases} \sum_{l}\exp(\lambda_{l}(t - t'))/N_{l} & \text{if } t > t', \\ 0 & \text{if } t < t'. \end{cases} \qquad\qquad (C.14)$$

For the case of an inhomogeneous equation

$$\dot{x}(t) = -ax(t) + bx(t - \tau) + f(t) \qquad\qquad (C.15)$$

and the initial condition (eq.(C.2)) the solution is expanded in eigenmodes:

$$x(t + \theta) = \sum_{l}C_{l}(t)\exp(\lambda_{l}\theta), \quad -\tau \leq \theta \leq 0. \qquad\qquad (C.16)$$

To determine the coefficients $C_l(t)$ we use the equations (C.10) and (C.12)

$$C_l(t) = \int_{-\tau}^{0} V_l^*(\theta) x(t+\theta) d\theta$$

$$= \left[x(t) + b \int_{t-\tau}^{t} \exp(\lambda_l(t-\theta-\tau)) x(\theta) d\theta \right] / N. \qquad (C.17)$$

Taking the time derivative yields

$$\dot{C}_l(t) = \left[\dot{x}(t) + \frac{d}{dt} \left(b \int_{t-\tau}^{t} \exp(\lambda_l(t-\theta-\tau)) x(\theta) d\theta \right) \right] / N. \qquad (C.18)$$

Using

$$\frac{d}{dt} \left(\int_{a(t)}^{b(t)} f(x,t) dx \right) = \int_{a(t)}^{b(t)} \frac{d}{dt} f(x,t) + f(b(t),t) \frac{d}{dt} b(t) - f(a(t),t) \frac{d}{dt} a(t) \qquad (C.19)$$

one can write the second term as

$$\frac{d}{dt} \left[\left(b \int_{t-\tau}^{t} \exp(\lambda_l(t-\theta-\tau)) x(\theta) d\theta \right) \right]$$

$$= b \int_{t-\tau}^{t} \frac{d}{dt} \exp(\lambda_l(t-\theta-\tau)) x(\theta) d\theta + b \exp(-\lambda_l \tau) x(t) - b x(t-\tau). \qquad (C.20)$$

Therefore, we obtain

$$N \dot{C}_l(t) = -ax(t) + bx(t-\tau) + f(t) + b\lambda_l \int_{t-\tau}^{t} \exp(\lambda_l(t-\theta-\tau)) x(\theta) d\theta$$

$$+ b \exp(-\lambda_l \tau) x(t) - b x(t-\tau)$$

$$= (-a + b \exp(-\lambda_l \tau)) x(t) + f(t) + b\lambda_l \int_{t-\tau}^{t} \exp(\lambda_l(t-\theta-\tau)) x(\theta) d\theta. \qquad (C.21)$$

Now we use equation (C.4) and we end up with

$$\dot{C}_l(t) = \left(\lambda_l x(t) + b\lambda_l \int_{t-\tau}^{t} \exp(\lambda_l(t-\theta-\tau)) x(\theta) d\theta + f(t) \right) / N$$

$$= \lambda_l C_l(t) + \frac{f(t)}{N}. \qquad (C.22)$$

Using $C_l(0) = c_l$ as initial condition, we obtain by integration

$$C_l(t) = \exp(\lambda_l t) c_l + \int_{0}^{t} \exp(\lambda_l(t-t')) f(t') dt' / N_l \qquad (C.23)$$

Therefore, the solution for the inhomogeneous equation (C.15) is

$$
\begin{aligned}
x(t) &= \sum_l C_l(t) \\
&= \sum_l \left(\exp(\lambda_l t) c_l + \int_0^t \exp(\lambda_l (t - t')) f(t') dt' / N_l \right) \\
&= \sum_l \left[\left(\phi(0) + \int_{-\tau}^0 b \exp(-\lambda_l (\theta + \tau)) \phi(\theta) d\theta \right) \frac{\exp(\lambda_l t)}{N_l} \right. \\
&\qquad \left. + \int_0^t \exp(\lambda_l (t - t')) f(t') dt' / N_l \right] \\
&= T(t, 0)\phi(0) + \int_{-\tau}^0 T(t, \theta + \tau) b \phi(\theta) d\theta + \int_0^t T(t, t') f(t') dt', \quad (\text{C.24})
\end{aligned}
$$

where we used the equations (C.8), (C.12), and (C.14).

Appendix D

Calculation of the power spectral density

Here we want to show the detailed calculation of the power spectral density from section 3.2.2. We make use of the Wiener-Khinchin theorem by computing the Fourier transform of the autocorrelation function:

$$
\begin{aligned}
S_{xx}(\omega) &= \int_{-\infty}^{\infty} e^{i\omega t} \langle x(t)x(0)\rangle dt \\
&= \int_{0}^{\infty} e^{i\omega t} \langle x(t)x(0)\rangle dt + \int_{-\infty}^{0} e^{i\omega t} \langle x(t)x(0)\rangle dt \\
&= \int_{0}^{\infty} e^{i\omega t} \langle x(t)x(0)\rangle dt + \int_{0}^{\infty} e^{-i\omega t} \langle x(t)x(0)\rangle dt \\
&= \int_{0}^{\infty} 2\cos(\omega t) \langle x(t)x(0)\rangle dt \\
&= 2\mathrm{Re} \int_{0}^{\infty} e^{i\omega t} \langle x(t)x(0)\rangle dt,
\end{aligned}
\tag{D.1}
$$

where we used the property that the autocorrelation function is an even function. Now we can perform the integration directly by using the autocorrelation function

$$
\begin{aligned}
\langle x(t)x(0)\rangle &= 2D \sum_{l,l'} \mathrm{Re} \frac{\exp(\Lambda_l t)}{N_l N_{l'}^*(-\Lambda_l - \Lambda_{l'}^*)} \\
&= 2D \frac{1}{2} \sum_{l,l'} \left(\frac{\exp(\Lambda_l t)}{N_l N_{l'}^*(-\Lambda_l - \Lambda_{l'}^*)} + \frac{\exp(\Lambda_l^* t)}{N_l^* N_{l'}(-\Lambda_l^* - \Lambda_{l'})} \right)
\end{aligned}
\tag{D.2}
$$

and obtain the power spectral density after some rearranging

$$
\begin{aligned}
S_{xx}(\omega) &= 2D\frac{1}{2}2\mathrm{Re}\sum_{l,l'}\left[\frac{1}{-\Lambda_l - i\omega}\frac{1}{N_l N_{l'}^*(-\Lambda_l - \Lambda_{l'}^*)}\right.\\
&\quad\left. + \frac{1}{-\Lambda_l^* - i\omega}\frac{1}{N_l^* N_{l'}(-\Lambda_l^* - \Lambda_{l'})}\right]\\
&= 2D\frac{1}{2}\sum_{l,l'}\left[\frac{1}{-\Lambda_l - i\omega}\frac{1}{N_l N_{l'}^*(-\Lambda_l - \Lambda_{l'}^*)} + \frac{1}{-\Lambda_l^* + i\omega}\frac{1}{N_l^* N_{l'}(-\Lambda_l^* - \Lambda_{l'})}\right.\\
&\quad\left. + \frac{1}{-\Lambda_l^* - i\omega}\frac{1}{N_l^* N_{l'}(-\Lambda_l^* - \Lambda_{l'})} + \frac{1}{-\Lambda_l + i\omega}\frac{1}{N_l N_{l'}^*(-\Lambda_l - \Lambda_{l'}^*)}\right]\\
&= 2D\frac{1}{2}\sum_{l,l'}\left[\left(\frac{1}{-\Lambda_l - i\omega} + \frac{1}{-\Lambda_l + i\omega}\right)\frac{1}{N_l N_{l'}^*(-\Lambda_l - \Lambda_{l'}^*)}\right.\\
&\quad\left. + \left(\frac{1}{-\Lambda_l^* - i\omega} + \frac{1}{-\Lambda_l^* + i\omega}\right)\frac{1}{N_l^* N_{l'}(-\Lambda_l^* - \Lambda_{l'})}\right]\\
&= 2D\mathrm{Re}\sum_{l,l'}\frac{-2\Lambda_l}{\Lambda_l^2 + \omega^2}\frac{1}{N_l N_{l'}^*(-\Lambda_l - \Lambda_{l'}^*)},
\end{aligned}
\tag{D.3}
$$

where we used for the last step

$$
\left(\frac{1}{-\Lambda_l - i\omega} + \frac{1}{-\Lambda_l + i\omega}\right) = \frac{-2\Lambda_l}{(-\Lambda_l - i\omega)(-\Lambda_l + i\omega)} = \frac{-2\Lambda_l}{\Lambda_l^2 + \omega^2}.
\tag{D.4}
$$

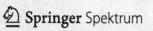